The Role of Plankton in Freshwater and Marine Ecology

Edited by Leonel Pereira

Published in London, United Kingdom

The Role of Plankton in Freshwater and Marine Ecology
http://dx.doi.org/10.5772/intechopen.110972
Edited by Leonel Pereira

Contributors
Adel Naseeb, Ahmad Al-Khayat, Ahmad J. Al-Shemmari, Alexandra Guo, Alexis Saldivar, Amina Amrani, Amira Rekik, Ana Bianca Pavel, Barbara Saucedo, Carmen Postolache, Chaima Aounallah, Cristal Zuniga, Daniel Norena-Caro, Darshini Parekh, Diego Martinez, Fateh Sedrati, Gabriela Canto-Encalada, Habib Ayadi, Hadjer Zaidi, Hanan Al-Adeelah, Hichem Nasri, Ila Peeler, Irene Gallego, Jannet Elloumi, Jenna Armstrong, Kayleen Snyder, Leonel Pereira, Loubna Ghodbane, Maaref Hanen, Marc Pagno, Maria Iasmina Moza, Mason Raymond, Mohammad Ali, Mohammad Boarki, Moshe Gophen, Nassima Beldjoudi, Neila Annabi-Trabelsi, Shaila Prasad, Shelly Bouyea, Timothy B. Mihuc, Wassim Guermazi, Zachary Cutter, Zakia Hammoum, Zineb Labidi

First published in London, United Kingdom, 2025 by IntechOpen
IntechOpen is the global imprint of INTECHOPEN LIMITED, registered in England and Wales, registration number: 11086078, 167-169 Great Portland Street, London, W1W 5PF, United Kingdom

For EU product safety concerns: IN TECH d.o.o., Prolaz Marije Krucifikse Kozulić 3, 51000 Rijeka, Croatia, info@intechopen.com or visit our website at intechopen.com.

British Library Cataloguing-in-Publication Data
A catalogue record for this book is available from the British Library

The Role of Plankton in Freshwater and Marine Ecology
Edited by Leonel Pereira
p. cm.
Print ISBN 978-0-85466-546-4
Online ISBN 978-0-85466-545-7
eBook (PDF) ISBN 978-0-85466-664-5

Meet the editor

Leonel Pereira has a degree in Biology (Scientific branch) and a Ph.D. in Biology (specializing in Cell Biology) from the Faculty of Science and Technology of the University of Coimbra in Portugal, where he is currently an Associate Professor with Habilitation. He also teaches at this university and is a CFE - Center for Functional Ecology: Science for People & Planet researcher. He is currently the Master's Degree Coordinator in Biodiversity and Plant Biotechnology, the Coordinator of LAM - Marine Algae Lab, and the Coordinator of ACOI - Coimbra Collection of Algae. His interests focus mainly on marine biodiversity (algae), marine biotechnology (bioactive compounds from macroalgae), and marine ecology (environmental assessment). Since 2008, he has been the author and editor of the electronic publication MACOI – Portuguese Seaweeds Website (http://www. flordeutopia. pt/macoi/). He is the author of more than 20 books and more than 80 book chapters. So far, he has published more than 170 scientific articles in international journals and given more than a hundred lectures and oral communications at various national and international scientific events. In 1998, he was awarded the Francisco de Holanda Prize (Honorable Mention) and, more recently, the King Carlos Sea Prize (18th edition). He is part of "The World's Top 2% Scientists" list for 2021, 2022, and 2023, as released by Stanford University and Elsevier.

Contents

Preface

Plankton are fundamental components of aquatic ecosystems, serving as the primary producers and forming the aquatic food web base. These microorganisms significantly influence global biogeochemical cycles, climate regulation, and biodiversity maintenance. Despite their microscopic size, plankton is critical in maintaining ecological equilibrium across freshwater and marine environments.

This volume aims to provide a comprehensive examination of the biology, ecological functions, and environmental interactions of plankton. It also explores their potential applications in biotechnology, sustainability, and environmental management. The content is organized into three sections, each focusing on a crucial aspect of plankton research.

Section 1, "Introduction", introduces the ecological significance of plankton in aquatic systems. The initial chapter by Leonel Pereira presents an overview of the diverse forms and ecological roles of plankton, providing a foundation for subsequent detailed discussions. Moshe Gophen investigates the complex interactions within the Lake Kinneret ecosystem, highlighting the relationships between biotic and abiotic factors. Historical analysis by Timothy B. Mihuc and colleagues examines 50 years of phytoplankton community changes in Lake Champlain, revealing the long-term effects of environmental variations. Maria Iasmina Moza and her team analyze the influence of environmental factors on cyanobacteria genotypes and toxin production. Additionally, Hichem Nasri and collaborators study the impact of physicochemical parameters on the growth and toxicity of *Microcystis sp.*, a common cyanobacterium associated with harmful algal blooms.

Section 2, "Marine and Coastal Ecosystem Studies", focuses on plankton dynamics in marine environments. Amira Rekik and her team explore the seasonal variability of zooplankton communities in the El Bibane Lagoon, emphasizing the impact of environmental factors on community structure and species interactions. This study underscores the sensitivity of marine planktonic communities to environmental fluctuations and their potential as indicators of ecosystem health.

Section 3, "Innovations and Applications", addresses advancements in plankton research and their practical applications. Barbara Saucedo and her team discuss the use of mathematical models to understand the metabolic pathways of photosynthetic plankton, which have implications for optimizing plankton cultivation for biotechnological applications. Irene Gallego concludes the book by examining biocompounds derived from freshwater and marine phytoplankton. These natural products, including pigments, antioxidants, and biofuels, offer significant potential for commercial and pharmaceutical applications, positioning plankton as a key resource for sustainable innovation.

This volume represents a collaborative effort by leading researchers worldwide, contributing diverse insights into plankton research. The chapters collectively highlight the ecological, environmental, and economic importance of plankton. By integrating fundamental research with applied sciences, this book aims to inspire new directions in plankton research and underscore their role in addressing global challenges such as climate change and food security.

The volume is intended as a valuable resource for scientists, students, and practitioners in aquatic ecology, environmental management, and biotechnology. It is hoped that readers will gain a deeper understanding of plankton and their potential to contribute to a sustainable future.

Leonel Pereira
CFE—Centre for Functional Ecology: Science for People & Planet,
Marine Resources, Conservation and Technology,
Marine Algae Lab,
Department of Life Sciences,
University of Coimbra,
Coimbra, Portugal

Section 1

Introduction

Chapter 1

The Unsung Heroes of Aquatic Ecosystems: The Vital Roles of Plankton

Leonel Pereira

Abstract

Plankton, a diverse group of organisms suspended in freshwater and marine ecosystems, plays a crucial role in aquatic environments. They are categorized into phytoplankton, zooplankton, and bacterioplankton, each contributing uniquely to ecological processes. Phytoplankton, as primary producers, drives photosynthesis and oxygen production, forming the base of aquatic food webs. Zooplankton, as primary consumers, link primary producers to higher trophic levels and aid in nutrient recycling. Bacterioplankton is essential for decomposing organic material and mediating biogeochemical cycles. Plankton distribution is influenced by environmental factors such as light, nutrients, temperature, and salinity, with distinct communities in freshwater and marine habitats. Understanding plankton dynamics is vital for appreciating their role in global biogeochemical cycles and ecosystem health.

Keywords: symbiosis, competition, aquatic ecosystems, biogeochemical cycles, blooms

1. Introduction

Plankton is a diverse group of organisms that live suspended in the water column of freshwater and marine ecosystems [1]. They are primarily defined by their limited ability to swim against currents, relying on water movements for their distribution. Despite their microscopic size, plankton plays a crucial role in aquatic ecosystems by forming the base of the food web and driving global biogeochemical cycles [2, 3].

Plankton can be broadly classified into three main categories based on their biological characteristics and ecological roles: phytoplankton, zooplankton, and bacterioplankton [1]. Phytoplankton are photosynthetic microorganisms, often referred to as the "grass of the sea". They use sunlight to convert carbon dioxide and water into oxygen and organic compounds, contributing significantly to global oxygen production [4]. Examples include diatoms (Bacillariophyceae), dinoflagellates (Dinoflagellata), blue-green algae (Cyanobacteria), and green algae (Chlorophyta) [5]. Phytoplankton forms the foundation of aquatic food webs, supporting herbivorous zooplankton and higher trophic levels [6]. Additionally, they play a key role in carbon sequestration through the biological pump, where carbon is transferred from the ocean's surface to its depths [7].

1.1 Zooplankton

Zooplankton, on the other hand, are heterotrophic organisms that feed on phytoplankton, other zooplankton, or organic detritus. Although they exhibit some mobility, they remain largely at the mercy of water currents. Common examples include copepods (Copepoda), krill (Crustacea), jellyfish larvae (Cnidaria), and protozoa such as ciliates (Ciliophora) and Foraminifera. Zooplankton serve as primary consumers, linking the energy generated by primary producers to higher trophic levels like fish, birds, and marine mammals. They also contribute to nutrient recycling by breaking down organic material. Their sensitivity to environmental changes makes them valuable indicators of water quality and ecosystem health [8, 9].

1.2 Bacterioplankton

Bacterioplankton, which includes Bacteria and Archaea in planktonic form, are a vital component of the microbial loop in aquatic systems. Examples range from heterotrophic bacteria like *Vibrio* species to nitrifying bacteria and methanogenic Archaea [10]. These microorganisms play a crucial role in decomposing organic material and recycling essential nutrients such as nitrogen and phosphorus back into the ecosystem. They also mediate key biogeochemical processes, including nitrogen fixation, denitrification, and sulfur cycling. Bacterioplankton often interacts with phytoplankton, competing for nutrients or forming mutualistic relationships, such as providing essential vitamins [11].

1.3 Plankton size

Plankton are also categorized based on size, life cycle, and habitat. Size-based classifications range from picoplankton, less than 2 μm in diameter (e.g., Cyanobacteria), to macroplankton, larger than 200 μm (e.g., jellyfish larvae) [1, 12]. In terms of life cycle, holoplankton spends their entire lives as plankton, while meroplankton are only planktonic during certain life stages, such as the larval phase of starfish (Asterozoa) [13, 14]. Habitat distinctions separate freshwater plankton, found in lakes, rivers, and ponds, from marine plankton, which inhabit oceans, seas, and estuaries [1].

This diverse classification underscores the ecological importance of plankton. By driving primary production, recycling nutrients, and supporting aquatic food webs, plankton sustain biodiversity and ecosystem health. Their roles as indicators of environmental change and contributors to global biogeochemical processes highlight their foundational position in the study of aquatic ecology [15]. Plankton are found in nearly all aquatic environments, with their distribution shaped by a combination of environmental factors such as light availability, nutrient levels, temperature, and salinity. Their presence spans both freshwater and marine ecosystems, with distinct communities adapted to each habitat [16].

In freshwater ecosystems, plankton inhabit lakes, rivers, ponds, reservoirs, and wetlands. The distribution and abundance of freshwater plankton are influenced by the size and depth of the water body, flow dynamics, and nutrient availability [17].

For instance, in lakes, phytoplankton populations often peak in the upper layers (epilimnion), where sunlight penetrates, while zooplankton follow the vertical migrations of their prey [18]. In fast-flowing rivers, plankton populations may be less stable due to turbulence and limited nutrient retention [19]. Eutrophic lakes and reservoirs, which are rich in nutrients, often support high densities of Cyanobacteria and green algae, while oligotrophic (nutrient-poor) lakes are dominated by smaller picoplankton and diatoms [20].

1.4 Marine ecosystems plankton

In marine ecosystems, plankton are distributed across vast oceanic regions, from shallow coastal waters to the open ocean and even into the deep-sea environment [2]. Coastal areas and estuaries, rich in nutrients from terrestrial inputs and upwelling processes, often harbor abundant and diverse plankton communities. These regions are dominated by larger phytoplankton like diatoms and dinoflagellates, which support productive food webs, including commercially significant fisheries [21]. In contrast, open ocean (pelagic) regions, where nutrients are limited, are dominated by smaller phytoplankton such as Cyanobacteria and nanoplankton, which are adapted to low-nutrient conditions [22]. Zooplankton distributions in marine systems often follow similar patterns, with species composition and abundance influenced by the availability of phytoplankton and physical factors like currents and water temperature [23].

Salinity plays a critical role in differentiating plankton communities between freshwater and marine systems. While marine phytoplankton like diatoms and dinoflagellates thrive in saline environments, freshwater systems are often dominated by green algae and Cyanobacteria, which are adapted to lower salinity [24]. Transitional environments such as estuaries support unique plankton communities that can tolerate wide salinity fluctuations, with organisms like euryhaline copepods bridging the gap between freshwater and marine species [25].

Overall, plankton distribution reflects the dynamic interplay of abiotic factors, such as nutrient availability and hydrodynamics, and biotic factors, including predation and competition. Understanding these patterns is crucial for appreciating the ecological roles plankton play in both local ecosystems and global processes, such as nutrient cycling and carbon sequestration [26]. Plankton are fundamental to aquatic food webs, serving as both primary producers and primary consumers, roles that are essential for maintaining ecosystem function and supporting biodiversity. Together, they drive the flow of energy and nutrients through freshwater and marine ecosystems, underpinning nearly all aquatic life [27].

As primary producers, phytoplankton performs photosynthesis, using sunlight, carbon dioxide, and water to produce organic matter and release oxygen. These microscopic, plant-like organisms are responsible for nearly half of the Earth's oxygen production, playing a critical role in sustaining life on the planet [28]. Phytoplankton forms the base of aquatic food webs by generating the biomass that supports herbivorous zooplankton, which then sustains higher trophic levels such as fish, marine mammals, and birds. In regions with abundant nutrients, such as coastal upwelling zones and estuaries, phytoplankton blooms drive high biological productivity, supporting diverse ecosystems and economically important fisheries [29].

As primary consumers, zooplankton feed on phytoplankton, bacteria, and organic detritus, converting these primary resources into forms of energy and biomass that can be utilized by larger predators [30]. Herbivorous zooplankton, like copepods and cladocerans (Cladocera), act as a vital intermediary, transferring energy from microscopic producers to fish, crustaceans, and other carnivores [8]. Carnivorous zooplankton, such as chaetognaths (arrow worms – Chaetognatha) and jellyfish larvae, consume smaller zooplankton, ensuring the energy flow continues through the food web [23]. These feeding relationships maintain ecosystem balance and enable the cycling of nutrients across different trophic levels [31].

Beyond their direct contributions to food webs, plankton plays a broader role in ecosystem processes [32]. Phytoplankton is central to the biological pump, a mechanism by which carbon is sequestered from the atmosphere to the deep ocean. By absorbing carbon dioxide during photosynthesis and contributing organic material to sinking particles, they help mitigate global climate change [7]. Zooplankton further enhance this process by producing fecal pellets that sink rapidly, transferring carbon and nutrients to deeper layers of the water column. This coupling of production and consumption is vital for regulating the Earth's carbon cycle and supporting nutrient recycling in aquatic environments [33].

Plankton's dual roles as primary producers and consumers are indispensable to aquatic ecosystems. They sustain food web dynamics, drive essential biogeochemical cycles, and support the productivity and stability of freshwater and marine systems. Without plankton, aquatic ecosystems would collapse, disrupting the balance of life both within and beyond these habitats [34].

2. Ecological roles of plankton

Plankton, particularly phytoplankton, plays a crucial ecological role as the primary producers in aquatic ecosystems, driving primary production and contributing significantly to oxygen generation. These microscopic, photosynthetic organisms form the foundation of food webs and are essential for maintaining ecological balance in both freshwater and marine environments [1].

2.1 Phytoplankton

Phytoplankton utilizes sunlight, carbon dioxide, and water to perform photosynthesis, producing organic compounds that serve as the primary source of energy for aquatic food webs. This process also releases oxygen as a byproduct, making phytoplankton major contributors to the global oxygen supply. It is estimated that phytoplankton is responsible for nearly 50% of the Earth's annual oxygen production, rivaling terrestrial forests in their capacity to sustain life. This oxygen is critical not only for aquatic organisms but also for maintaining atmospheric balance and supporting terrestrial and marine life alike [3].

In terms of primary production, phytoplankton are remarkably efficient at converting solar energy into biomass, providing the energy necessary for herbivorous zooplankton, which then feed higher trophic levels [35]. This role is especially pronounced in nutrient-rich regions such as coastal upwelling zones, estuaries, and polar seas, where phytoplankton blooms can lead to extraordinarily high levels of biological productivity [36]. In nutrient-poor areas like the open ocean, smaller phytoplankton, such as Cyanobacteria and picoplankton, dominate and sustain food webs adapted to oligotrophic conditions [37].

Phytoplankton also plays a key role in the global carbon cycle through their involvement in the biological pump. During photosynthesis, they absorb carbon dioxide from the atmosphere and incorporate it into organic material. Some of this material sinks to the ocean depths as particulate organic carbon, effectively sequestering carbon and mitigating climate change. This dual contribution to both oxygen generation and carbon cycling underscores the central role of phytoplankton in regulating Earth's climate systems [38]. The contribution of plankton to primary production and oxygen generation is a cornerstone of aquatic and global ecological processes. By driving food web dynamics, supporting biodiversity, and influencing atmospheric and oceanic systems, plankton underscore the interconnectedness of life on Earth. Their role highlights the importance of conserving aquatic ecosystems and understanding the factors that influence plankton dynamics in a changing environment [39].

Plankton are integral to nutrient cycling in aquatic ecosystems, driving the transformation and redistribution of key elements such as carbon, nitrogen, and phosphorus. These cycles are essential for maintaining ecosystem productivity, supporting biodiversity, and regulating global biogeochemical processes [40].

2.2 Carbon cycle

Phytoplankton play a central role in the carbon cycle through photosynthesis, where they absorb carbon dioxide (CO_2) from the atmosphere and convert it into organic compounds. This process not only supports their growth but also forms the basis of the aquatic food web. When plankton is consumed or dies, the carbon stored in their bodies can follow various pathways [41]. Some of it is recycled within the upper water column through microbial decomposition, while a portion sinks to deeper waters as organic detritus or fecal pellets produced by zooplankton. This sinking carbon contributes to the biological pump, a critical mechanism for sequestering carbon in the deep ocean, where it can remain for centuries. By reducing atmospheric CO_2 levels, plankton play a significant role in mitigating climate change [42].

2.3 Nitrogen cycle

Plankton are also key players in the nitrogen cycle, particularly through their roles in nitrogen fixation, assimilation, and regeneration. Certain phytoplankton, such as Cyanobacteria (e.g., *Trichodesmium*) (**Figure 1**), are capable of nitrogen fixation, converting atmospheric nitrogen (N_2) into bioavailable forms like ammonium

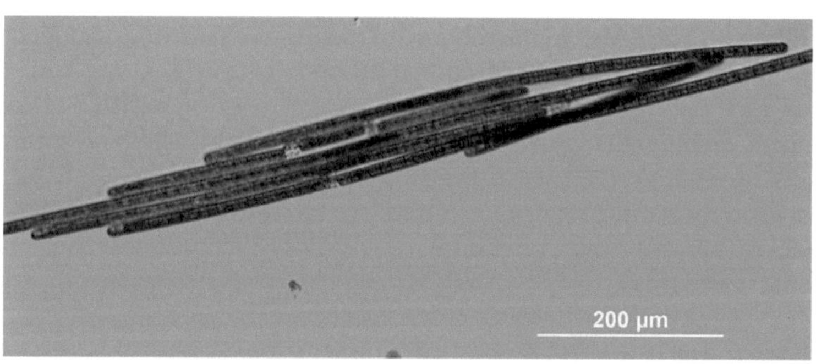

Figure 1.
Picture of Trichodesmium *sp. by Prof. Ondøej Práil, Institute of Microbiology, Czech Academy of Sciences, Czech Republic. Creative Commons Attribution-Share Alike 4.0 International license.*

(NH_4^+) [43]. This process provides an essential source of nitrogen in nutrient-poor (oligotrophic) regions, supporting primary production. Zooplankton and bacterioplankton further influence the nitrogen cycle by recycling nitrogen compounds. When zooplankton consumes phytoplankton or detritus, they excrete waste products like ammonium, which can be taken up by phytoplankton for growth [44]. Bacterioplankton contributes to processes such as nitrification (oxidation of ammonium to nitrate) and denitrification (conversion of nitrate to nitrogen gas), completing the nitrogen cycle and regulating nitrogen availability [45].

2.4 Phosphorus cycle

Phosphorus, primarily in the form of phosphate (PO_4^{3-}), is another essential nutrient cycled by plankton. Phytoplankton assimilates phosphate from the water column to build cellular components like nucleic acids and membranes. As plankton dies or is consumed, phosphorus is released back into the water through decomposition or excretion by zooplankton and bacteria [46]. Bacterioplankton plays a significant role in mineralizing organic phosphorus, converting it back into inorganic phosphate, which can be reused by primary producers. This regenerative loop ensures the availability of phosphorus, a limiting nutrient in many aquatic systems [47].

2.5 Plankton in nutrient hotspots

In regions like upwelling zones, estuaries, and areas influenced by riverine inputs, plankton activity accelerates nutrient cycling. These hotspots often experience rapid turnover of carbon, nitrogen, and phosphorus, leading to high productivity and supporting complex food webs. Conversely, in nutrient-poor regions, plankton adapts to efficiently utilize scarce resources, maintaining ecosystem function even under limiting conditions [48].

2.6 Global significance

Plankton-mediated nutrient cycling not only supports local ecosystems but also has far-reaching implications for global environmental processes. By driving the movement of carbon, nitrogen, and phosphorus through aquatic ecosystems and into long-term reservoirs, plankton helps regulate climate, sustain fisheries, and maintain the balance of life in both freshwater and marine environments [49]. Plankton is indispensable to nutrient cycling, ensuring the availability and flow of essential elements within and beyond aquatic ecosystems. Their roles highlight the interconnected nature of nutrient dynamics and underscore their importance in sustaining life on Earth [50].

Plankton serves as highly effective indicators of ecosystem health and environmental changes due to their sensitivity to physical, chemical, and biological factors in aquatic ecosystems. As the base of the food web and direct participants in biogeochemical cycles, plankton responds rapidly to alterations in their environment, making them invaluable for monitoring ecological stability and detecting emerging environmental issues [51].

2.7 Sensitivity to environmental conditions

Plankton communities are highly influenced by variations in water temperature, light availability, salinity, nutrient levels, and pollution. For example, changes in

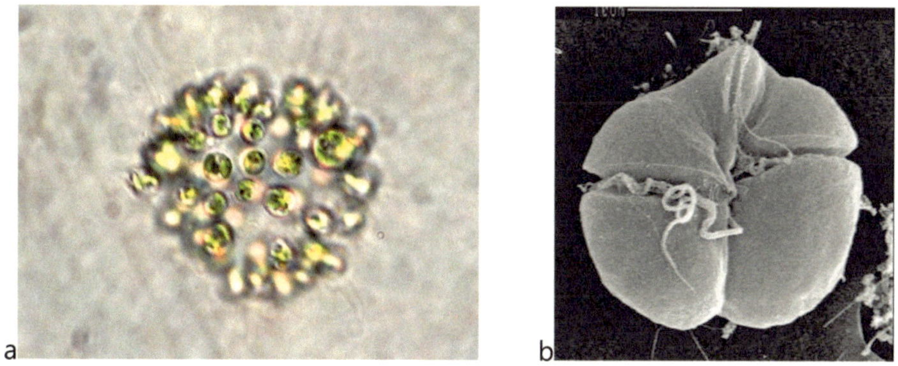

Figure 2.
a. Microcystis aeruginosa, *by Kristian Peters http://www.korseby.net/outer/flora/algae/index.html – Own work. Creative Commons Attribution-Share Alike 3.0 International license. b.* Karenia brevis, *by Florida Fish and Wildlife Conservation Commission. U.S. Fish and Wildlife Service. Image is in the public domain.*

nutrient availability, such as eutrophication from agricultural runoff, can lead to phytoplankton blooms dominated by harmful algal species like *Microcystis* (Cyanobacteria) (a) or *Karenia brevis* (Dinoflagellata) (b) (**Figure 2**). These blooms may produce toxins that harm aquatic organisms and humans, signaling declining water quality. Conversely, a decrease in phytoplankton abundance or diversity may indicate nutrient depletion, pollution stress, or unfavorable physical conditions such as reduced light penetration [52, 53].

Zooplankton, which feed on phytoplankton and other organic material, are also sensitive to environmental changes. Shifts in zooplankton species composition can reflect broader ecological disruptions, such as changes in predator populations, water quality, or habitat structure. For instance, a decline in copepod populations, which are a crucial food source for fish larvae, may foreshadow declines in higher trophic levels [54–56].

2.8 Indicators of climate change

Plankton dynamics are closely tied to water temperature, making them excellent indicators of climate change. Warming waters can lead to changes in the timing and magnitude of phytoplankton blooms, with cascading effects on zooplankton and higher trophic levels. Warmer temperatures may also favor the proliferation of harmful algal blooms and shifts in the composition of the plankton species, such as the dominance of smaller, warm-adapted phytoplankton species in place of larger diatoms [55]. In polar regions, planktonic shifts are among the first visible signs of climate-induced changes. For example, receding sea ice alters the timing of phytoplankton blooms, which can disrupt the tightly linked food web dynamics in these sensitive ecosystems. Similarly, acidification of oceans caused by increased CO_2 absorption can reduce calcification in species like coccolithophores and pteropods, indicating shifts in carbonate chemistry that affect marine biodiversity and ecosystem function [56].

2.9 Biodiversity and ecosystem monitoring

Plankton diversity and abundance are often used as metrics for assessing ecosystem health. High diversity typically indicates a stable and resilient ecosystem, while a decline in diversity or the dominance of a few opportunistic species may signal environmental stress or degradation. Long-term monitoring programs, such as those tracking phytoplankton and zooplankton populations in major lakes or oceans, provide valuable data for detecting trends in ecosystem health over time [57].

2.10 Applications in management and conservation

The use of plankton as ecological indicators supports effective management and conservation strategies. For instance, monitoring harmful algal blooms helps mitigate their impact on fisheries, tourism, and public health. Similarly, tracking zooplankton populations can inform the management of fish stocks, as these organisms are critical prey for commercially important species. Planktonic indicators also aid in assessing the success of restoration projects in degraded aquatic systems and detecting the impacts of pollution, overfishing, and habitat destruction [58, 59].

Plankton are sensitive, dynamic, and integral components of aquatic ecosystems, making them invaluable indicators of ecosystem health and environmental changes. By analyzing changes in their composition, abundance, and diversity, scientists and policymakers can better understand the impacts of human activities and natural processes on aquatic ecosystems, enabling informed decision-making to preserve ecological balance and sustain biodiversity [60, 61].

2.11 Blooms and their impacts

Bloom Characteristics and Causes Blooms: often referred to as algal blooms, are rapid increases in the population of algae or phytoplankton in aquatic environments. These events are typically triggered by a combination of factors, including nutrient enrichment (often from agricultural runoff), favorable light conditions, and stable water temperatures. Blooms can occur in both freshwater and marine ecosystems and are characterized by the discoloration of water, often turning green, red, or brown depending on the type of algae involved. While some blooms are harmless, others, known as harmful algal blooms (HABs), can produce toxins that pose risks to aquatic life, animals, and humans [55].

Ecological and Fisheries Impacts: the effects of blooms on ecosystems can be profound. In many cases, they lead to hypoxia (low oxygen levels) as the dense algae die off and decompose, consuming oxygen in the process. This can result in "dead zones" where aquatic life cannot survive. Additionally, toxins produced by HABs can kill fish, shellfish, and other marine organisms, disrupting food webs and diminishing biodiversity. Fisheries, particularly those reliant on shellfish, can suffer significant economic losses due to bloom events. The accumulation of toxins in shellfish can make them unsafe for consumption, leading to closure of fishing areas and affecting local economies dependent on these resources [21, 29].

Human Health and Activities: blooms also have a direct impact on human activities and health. Recreational waters can become unsafe for swimming or other activities, leading to beach closures and a decline in tourism. HABs pose health risks to humans through the consumption of contaminated seafood, causing illnesses such as shellfish poisoning. The economic implications extend beyond the seafood industry to public health costs and the loss of income from tourism and recreational activities. Effective monitoring and management strategies are essential to mitigate the impacts of blooms, protect ecosystems, and safeguard human health [29].

3. Plankton diversity and adaptations

Plankton exhibits remarkable diversity and a range of adaptations that enable them to thrive in both freshwater and marine environments. This diversity encompasses various *taxa*, functional roles, and ecological strategies, reflecting the unique challenges and opportunities presented by these habitats [62, 63]. Plankton can be broadly classified into three main groups: phytoplankton, zooplankton, and bacterioplankton. Phytoplankton, such as diatoms, dinoflagellates, green algae, Cyanobacteria, and coccolithophores, are photosynthetic organisms that form the base of aquatic food webs and drive primary production [64]. Zooplankton, including protozoans, crustaceans like copepods and cladocerans, and other small heterotrophic organisms, act as primary consumers, linking primary producers to higher trophic levels. Bacterioplankton, including bacteria and archaea, play essential roles in nutrient cycling, such as nitrogen fixation and carbon recycling [65].

3.1 Freshwater plankton

Despite their shared ecological roles, plankton communities differ significantly between freshwater and marine ecosystems due to variations in salinity, nutrient availability, and physical conditions. In freshwater environments, plankton faces challenges like fluctuating temperatures, variable nutrient concentrations, and low salinity [66]. Adaptations such as osmoregulation help maintain osmotic balance, while some phytoplankton, like green algae and Cyanobacteria, are equipped to rapidly exploit nutrient pulses from terrestrial inputs by forming blooms [67]. Freshwater zooplankton, such as cladocerans, often produce resting eggs during adverse conditions, ensuring survival and repopulation when environmental conditions improve. Some freshwater plankton are mixotrophic, combining photosynthesis with heterotrophy to survive in nutrient-poor conditions by consuming organic matter [68].

3.2 Marine plankton

Marine plankton, in contrast, must cope with high salinity, vast spatial scales, and strong vertical gradients in light and nutrients. Many marine plankton exhibit adaptations for buoyancy regulation, such as diatoms with siliceous frustules that reduce sinking or zooplankton like copepods that use lipid reserves for buoyancy. Marine species are also osmotically adapted to high salinity, with cellular mechanisms that prevent dehydration. In nutrient-poor open ocean regions, smaller planktons like picoplankton and nanoplankton dominate due to their high surface area-to-volume ratios, which enhance nutrient uptake efficiency [69]. Additionally, many zooplankton, such as copepods and krill, perform diel vertical migrations, ascending to surface waters at night to feed and descending to deeper waters during the day to avoid predators. Symbiotic relationships are also common in marine environments, such as mutualistic associations between dinoflagellates and corals or dinoflagellates hosting photosynthetic endosymbionts [70].

3.3 Adaptation and diversity of plankton in freshwater versus marine environments

The differences between freshwater and marine plankton communities are often shaped by their respective environmental conditions. Freshwater systems are typically dominated by green algae, Cyanobacteria, and smaller zooplankton like rotifers, whereas marine systems, especially open oceans, support diverse phytoplankton communities, including diatoms, dinoflagellates, and coccolithophores, alongside larger zooplankton like copepods, chaetognaths, and krill [71]. These adaptations not only ensure the survival of plankton in their respective habitats but also highlight their ecological significance as energy transfer, nutrient cycling, and global biogeochemical processes. Their diversity and resilience underscore their vital role in maintaining ecosystem stability and the interconnected n plankton exhibits a wide range of responses to abiotic factors such as temperature, light, salinity, and nutrients, reflecting their adaptability to diverse aquatic environments. These factors profoundly influence their physiology, distribution, growth, and ecological roles [72].

3.3.1 Temperature

Temperature is a critical determinant of planktonic activity and community composition. Phytoplankton growth rates typically increase with temperature up to an optimal range, beyond which thermal stress can inhibit photosynthesis and metabolism. Warm water species, such as Cyanobacteria, often dominate tropical and temperate regions, while cold-adapted species, like diatoms, thrive in polar and subpolar waters [73]. Temperature also affects zooplankton development and reproductive cycles. For instance, warmer temperatures can accelerate the life cycles of copepods, leading to shifts in predator-prey dynamics. Seasonal temperature changes drive phenomena like spring phytoplankton blooms in temperate zones, as rising temperatures stratify the water column and enhance light availability for photosynthesis [74].

3.3.2 Light

Light availability is essential for phytoplankton, as it drives photosynthesis. Light penetration in the water column varies with depth, turbidity, and season, shaping the vertical distribution of plankton. Phytoplankton in surface layers maximize light capture, while deeper-dwelling species often possess pigments adapted to absorb low-intensity light at specific wavelengths [75, 76]. Zooplankton, though not directly dependent on light for energy, are influenced by diel light cycles, often exhibiting diel vertical migration to avoid predation during daylight hours. In polar regions, extreme seasonal variations in light availability profoundly affect plankton dynamics, with bursts of productivity during periods of continuous daylight and dormancy during prolonged darkness [77].

3.3.3 Salinity

Salinity is a defining factor in distinguishing freshwater from marine plankton. Freshwater plankton are adapted to low-salinity conditions, employing osmoregulatory mechanisms to prevent excessive water influx. In contrast, marine plankton maintains osmotic balance in high-salinity environments through cellular ion regulation. Euryhaline species, capable of tolerating a wide range of salinities, are often found in estuaries and brackish waters, where salinity fluctuates due to tidal mixing and freshwater inputs. Changes in salinity, such as those caused by freshwater inflows, evaporation, or climate-induced alterations, can shift plankton community composition and productivity [78, 79].

Figure 3.
Harmful Cyanobacteria Bloom in the Banter Lake (Germany). The slimy mass was concentrated at the Grodendamm, by wind drift. By Ein Dahmer – Own work. Creative Commons Attribution-Share Alike 4.0 International license.

3.3.4 Nutrients

Nutrients are a primary driver of plankton growth and distribution. Phytoplankton requires macronutrients like nitrogen (N), phosphorus (P), and silica (Si) and micronutrients such as iron (Fe) for cellular processes. Nutrient availability often limits primary production, with nitrogen commonly limiting in marine systems and phosphorus in freshwater ecosystems [80]. Upwelling zones, where nutrient-rich deep water rises to the surface, support high plankton productivity, while nutrient-depleted (oligotrophic) regions like open oceans favor smaller phytoplankton adapted to low-nutrient conditions [76, 81]. Nutrient imbalances or surpluses, often caused by anthropogenic inputs, can trigger harmful algal blooms (HABs) (**Figure 3**), characterized by excessive growth of toxic phytoplankton species [82]. Zooplankton indirectly responds to nutrient levels through their dependence on phytoplankton as a food source, with their populations peaking during or shortly after phytoplankton blooms [83].

The responses of plankton to abiotic factors like temperature, light, salinity, and nutrients demonstrate their adaptability to varying environmental conditions. These factors not only regulate plankton physiology and behavior but also shape their ecological interactions, distribution patterns, and contributions to ecosystem processes. Understanding these responses is critical for predicting plankton dynamics in the face of environmental changes and managing aquatic ecosystem health [84].

4. Interactions within ecosystems

Plankton plays pivotal roles in aquatic food webs, forming the foundation of trophic interactions and driving the transfer of energy and nutrients across ecosystems.

These interactions are highly dynamic and complex, involving both direct feeding relationships and indirect effects that shape the structure and functioning of aquatic communities [85].

At the base of the food web, phytoplankton serves as primary producers, using sunlight and nutrients to generate organic matter through photosynthesis. This production supports a diverse array of organisms, from microscopic zooplankton to large marine mammals like whales [86]. Zooplankton, in turn, are the primary consumers, feeding on phytoplankton and forming a critical link between primary producers and higher trophic levels. For instance, copepods, a dominant group of zooplankton, consume phytoplankton and are themselves preyed upon by fish larvae, small fish, and other secondary consumers. These interactions are the backbone of energy transfer in aquatic ecosystems [67, 87, 88].

Higher up in the food web, nekton (such as fish, squid, and marine mammals) and other predators rely on zooplankton as an essential food source. For example, krill, a type of zooplankton, is a keystone species in polar ecosystems, serving as a primary food source for predators like whales, seals, and penguins. Changes in krill abundance can have cascading effects on the entire food web, illustrating the critical role of plankton in maintaining ecological balance [89].

Planktonic trophic interactions also involve microbial food webs, which are particularly important in nutrient cycling and energy flow. Bacterioplankton and protozoans consume dissolved organic matter and detritus, recycling nutrients back into forms usable by phytoplankton. This microbial loop complements the traditional grazing food web, ensuring the efficient use of organic material and sustaining productivity, especially in nutrient-poor (oligotrophic) regions [90]. Beyond direct consumption, plankton interacts through indirect processes such as competition and facilitation. For example, blooms of one phytoplankton species can outcompete others for nutrients or light, altering the community composition. Conversely, the presence of certain planktonic species can enhance the survival of others through nutrient remineralization or mutualistic relationships, such as nitrogen-fixing Cyanobacteria enriching nitrogen-poor waters for other phytoplankton [91].

Environmental changes, such as temperature shifts, nutrient loading, and ocean acidification, can significantly affect trophic interactions. For instance, rising temperatures may favor smaller phytoplankton species, which could reduce the energy transfer efficiency to larger zooplankton and fish. Similarly, nutrient imbalances from agricultural runoff can promote harmful algal blooms, disrupting food web dynamics and causing declines in zooplankton and fish populations [92].

Trophic interactions involving plankton are the cornerstone of aquatic food webs, facilitating energy flow from primary producers to apex predators and ensuring ecosystem stability. The intricate connections between phytoplankton, zooplankton, and other organisms underscore the importance of plankton in supporting biodiversity, regulating biogeochemical cycles, and maintaining the productivity of freshwater and marine systems [93]. Understanding these dynamics is essential for predicting ecosystem responses to environmental changes and managing aquatic resources sustainably [94].

Planktonic organisms engage in a wide array of symbiotic and competitive relationships that play essential roles in maintaining ecosystem balance and driving ecological processes. These interactions, ranging from mutualism to parasitism, and from direct competition to resource partitioning, are critical for the survival and coexistence of diverse planktonic species in aquatic ecosystems [95, 96].

4.1 Symbiotic relationships

Symbiosis is common among plankton and includes mutualism, commensalism, and parasitism. In mutualistic relationships, both organisms benefit. One well-known example is the partnership between coral reef-building organisms and photosynthetic dinoflagellate (*Symbiodinium*) [97, 98]. Although primarily associated with coral, these dinoflagellates also form symbiotic relationships with other planktonic organisms, such as jellyfish and radiolarians, providing energy through photosynthesis in exchange for nutrients like nitrogen and phosphorus [99]. Another form of mutualism occurs within microbial plankton communities. Nitrogen-fixing Cyanobacteria, such as *Trichodesmium*, benefit other phytoplankton by converting atmospheric nitrogen into bioavailable forms, supporting productivity in nutrient-poor waters. Similarly, certain diatoms form symbioses with nitrogen-fixing Cyanobacteria, enhancing growth for both partners in environments where nitrogen is limited (**Figure 4**) [100].

Commensalism is also prevalent in planktonic ecosystems. For instance, some zooplankton species benefit from associating with larger, non-predatory organisms that offer protection or transport without being harmed or benefited in return. Certain bacteria attach to larger planktonic organisms, taking advantage of nutrient-rich microenvironments around their hosts without impacting them [65, 101]. Parasitic relationships among plankton include interactions where one organism benefits at the expense of the other. Parasitic fungi, such as chytrids (Chytridiomycota), infect phytoplankton like diatoms and Cyanobacteria, reducing their populations and influencing bloom dynamics. These infections can significantly alter community structure and nutrient cycling [102, 103].

4.2 Competitive relationships

Competition among plankton typically arises from the limited availability of resources, such as light, nutrients, or space. Phytoplankton competes intensely for

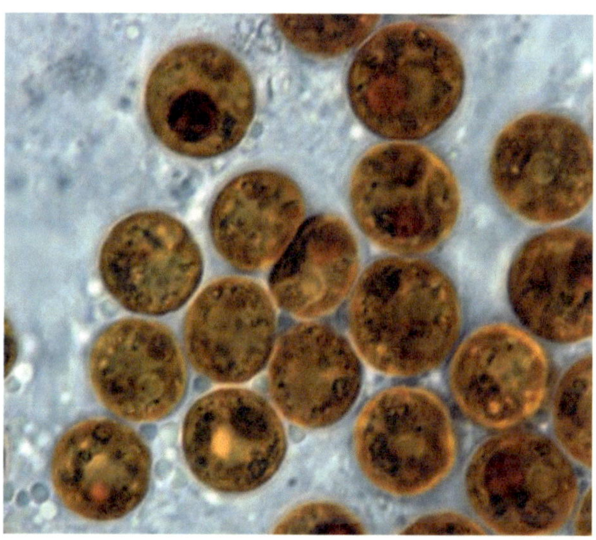

Figure 4.
Various specimens of the genus Symbiodinium, *by Allisonmlewis – Own work. Creative Commons Attribution-Share Alike 4.0 International license.*

light in the upper layers of the water column, with species that possess efficient light-harvesting pigments, like chlorophyll and accessory pigments, often outcompete with others under specific light conditions [104, 105]. For example, green algae dominate in nutrient-rich freshwater systems, while diatoms and Cyanobacteria can thrive in environments with different light intensities and nutrient profiles [106].

Nutrient competition is another major driver of planktonic dynamics. Phytoplankton species compete for essential nutrients like nitrogen, phosphorus, silica, and iron. Diatoms, for instance, require silica for their frustules, making them sensitive to silica availability. When silica is depleted, other phytoplankton, such as dinoflagellates or Cyanobacteria, may dominate. Similarly, blooms of certain species can deplete nutrients and suppress the growth of competitors, leading to shifts in community composition [107]. Allelopathy, a form of chemical competition, is also observed among plankton. Some species release chemicals that inhibit the growth of competitors. For instance, certain dinoflagellates produce toxins that can suppress the growth of co-occurring phytoplankton, providing a competitive advantage during blooms [108].

4.3 Balancing Symbiosis and competition

In many cases, planktonic organisms navigate a delicate balance between symbiosis and competition. For example, while nitrogen-fixing Cyanobacteria benefit other phytoplankton, they also compete for light and space. Similarly, zooplankton grazing can control phytoplankton populations, indirectly benefiting non-grazed phytoplankton species by reducing competition for nutrients. These interactions highlight the complex interplay between facilitation and competition in planktonic ecosystems [98, 109].

4.4 Ecological implications

The symbiotic and competitive relationships among plankton influence species diversity, population dynamics, and nutrient cycling. Symbioses enhance productivity in nutrient-poor regions, while competition determines community structure and succession. Together, these interactions ensure ecosystem resilience and adaptability to environmental changes [110]. The intricate web of symbiotic and competitive relationships among plankton underscores their ecological significance. Understanding these interactions provides valuable insights into ecosystem functioning, helping to predict responses to environmental stressors such as climate change, nutrient enrichment, and habitat alteration [111].

5. Plankton in global context

Plankton plays a vital role in global carbon sequestration and climate regulation, making them indispensable to Earth's biogeochemical cycles and climate systems. Their ability to absorb carbon dioxide (CO_2), produce oxygen, and drive nutrient cycling places them at the heart of processes that influence atmospheric CO_2 levels, oceanic carbon storage, and overall climate stability [112].

5.1 Phytoplankton and the biological carbon pump

Phytoplankton, the photosynthetic component of plankton, is critical in the biological carbon pump, a process by which carbon is transferred from the atmosphere to the ocean depths. These microscopic organisms use sunlight to convert CO_2 into organic matter through photosynthesis. This process not only contributes to nearly half of global primary production but also helps remove significant amounts of CO_2 from the atmosphere [38]. Once carbon is incorporated into phytoplankton biomass, it enters the marine food web. As phytoplankton is consumed by zooplankton and other organisms, a portion of the carbon is respired back into the water or atmosphere. However, some of the organic material sinks to deeper ocean layers as part of fecal pellets, dead organisms, or aggregates. This sequestration of carbon in the deep ocean can lock it away in centuries to millennia, reducing atmospheric CO_2 concentrations and mitigating the effects of climate change [113, 114].

5.2 Role of calcifying plankton

Certain planktonic organisms, such as coccolithophores, contribute to carbon cycling through the formation of calcium carbonate ($CaCO_3$) shells. While this process releases CO_2 into surface waters during shell formation, the sinking and long-term deposition of these shells in marine sediments also sequester carbon over geological timescales. This dual role underscores the complexity of Plankton's contributions to the global carbon budget [115, 116].

5.3 Zooplankton and carbon transport

Zooplankton are key players in carbon export through their grazing activities and diel vertical migration. By feeding phytoplankton near the surface and releasing waste products or dying in deeper waters, they enhance the downward flux of organic carbon. Vertical migration, where zooplankton moves to surface waters at night to feed and descends during the day, further aids in transporting carbon to the ocean's depths, facilitating its storage away from the atmosphere [117].

5.4 Bacterioplankton and microbial carbon processing

Bacterioplankton and other microbes are crucial for re-mineralizing organic matter, breaking it down into inorganic forms that can be reused by phytoplankton or transported deeper into the ocean. This microbial loop ensures efficient recycling of carbon and nutrients, supporting productivity and regulating the flow of carbon in marine ecosystems [118, 119].

5.5 Plankton and climate regulation

In addition to carbon sequestration, plankton influences climate regulation through the production of climate-active compounds. For instance, certain phytoplankton, such as *Gephyrocapsa huxleyi* (formerly *Emiliania huxleyi*) (Coccolithophyceae) (**Figure 5a**), produce dimethylsulfoniopropionate (DMSP) (**Figure 5b**), a precursor to dimethyl sulfide (DMS). DMS contributes to cloud formation by serving as cloud condensation nuclei, which reflects sunlight and cools the Earth's surface. This feedback mechanism links planktonic activity directly to atmospheric processes and climate modulation [120, 121].

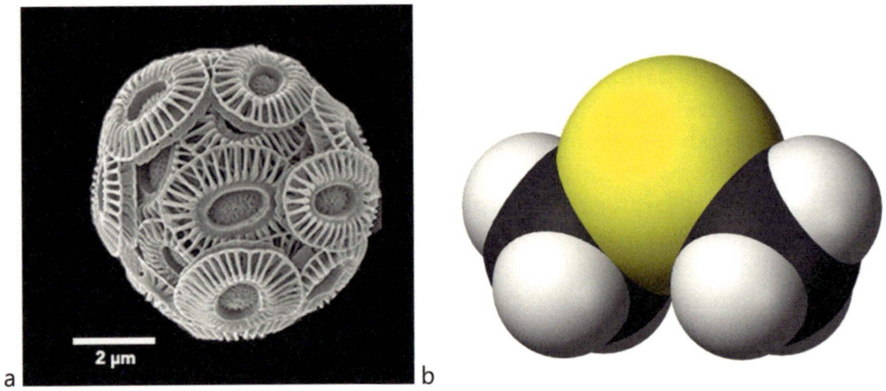

Figure 5.
a. Gephyrocapsa huxleyi, *by Dr. Jeremy Young, University College London – Extracted from this Commons file, CC BY-SA 4.0, https://commons.wikimedia.org/w/index.php?curid=109751103 b. dimethylsulfoniopropionate (DMSP), by Benjah-bmm27 - Own work, Public Domain, https://commons.wikimedia.org/w/index.php?curid=952850*

5.6 Threats to Plankton's role in carbon and climate regulation

Anthropogenic stressors, including ocean warming, acidification, and nutrient imbalances, threaten the efficiency of planktonic carbon sequestration. Rising temperatures can stratify ocean layers, reducing nutrient availability for phytoplankton and potentially lowering primary production [122]. Ocean acidification impairs calcifying organisms, such as coccolithophores and foraminifera, which are essential for long-term carbon storage in marine sediments [123]. Moreover, shifts in plankton community composition due to climate change and pollution may favor smaller, less efficient species for carbon export, diminishing the effectiveness of the biological carbon pump. Harmful algal blooms (HABs), often fueled by eutrophication, further disrupt planktonic processes, potentially releasing stored carbon back into the atmosphere [124].

5.7 Global implications

The contributions of plankton to carbon sequestration and climate regulation are pivotal in maintaining Earth's climate equilibrium. By absorbing large amounts of CO_2 and supporting long-term carbon storage, plankton mitigate the impacts of anthropogenic emissions. However, disruptions to their ecological roles could exacerbate climate change, emphasizing the need for global efforts to protect planktonic ecosystems and their functions [125]. Plankton are fundamental to carbon sequestration and climate regulation, driving processes that stabilize atmospheric CO_2 levels and influence Earth's climate. Their contributions underscore the interconnectedness of biological and physical systems in maintaining global environmental balance. Protecting these microscopic yet powerful organisms is critical for ensuring the health of the planet and its climate [126].

Plankton, with their remarkable diversity and biochemical capabilities, hold immense potential for biotechnological applications across various fields, including energy production, pharmaceuticals, and environmental management. Their rapid growth rates, adaptability to diverse environments, and ability to produce high-value biomolecules make them ideal candidates for sustainable and innovative technologies [127].

5.8 Biofuels and renewable energy

Phytoplankton, particularly microalgae, are gaining attention as a sustainable source of biofuels. These organisms efficiently convert sunlight and CO_2 into biomass, with some species accumulating high levels of lipids that can be processed into biodiesel. Unlike terrestrial biofuel crops, microalgae do not require arable land or freshwater for cultivation, making them a more environmentally friendly option [115]. Species such as *Chlorella* (Chlorophyta) and *Nannochloropsis* (Eustigmatophyceae) are extensively studied for biofuel production due to their high lipid content and rapid growth rates. Advances in genetic engineering have further enhanced lipid yield, improving the economic feasibility of algal biofuels. Additionally, some Cyanobacteria are being explored for biohydrogen production, a clean energy source, through their natural photosynthetic pathways [128, 129].

5.9 Pharmaceuticals and nutraceuticals

Plankton produces a vast array of bioactive compounds with significant pharmaceutical potential. For instance, diatoms and Cyanobacteria synthesize unique secondary metabolites with antibacterial, antiviral, and anticancer properties. *Spirulina/Arthrospira/Limnospira* sp. (**Figure 6**), blue-green algae, is widely regarded as a superfood due to its exceptional nutritional profile. Comprising 60–70% protein by dry weight, it ranks among the most protein-rich foods available. As a complete protein, *Spirulina/Arthrospira/Limnospira* provides all essential amino acids, making it especially valuable for vegetarians, vegans, and anyone looking to boost their protein intake. Beyond its protein content, *Spirulina/Arthrospira/Limnospira* are packed with vital vitamins and minerals, including B vitamins, iron, calcium, magnesium, and potassium. These nutrients play critical roles in supporting energy production, immune function, and overall health.

Certain dinoflagellates and Cyanobacteria produce toxins, such as saxitoxin and brevetoxin (**Figure 7**), which, while harmful in natural ecosystems, have been studied for their potential use in developing pain-relieving drugs or neuroprotective agents. Microalgae such as *Haematococcus lacustris* (Chlorophyta) produces astaxanthin, a powerful antioxidant with applications in treating oxidative stress-related diseases [129, 130]. Marine plankton, including some foraminifera and radiolarians, are also being investigated for their potential to produce biomineralized structures, which can be used in regenerative medicine, particularly in bone and dental repair [131, 132].

5.10 Nutritional applications

Plankton, especially microalgae, are a valuable source of omega-3 fatty acids (e.g., eicosapentaenoic acid [EPA] and docosahexaenoic acid [DHA]), which are essential for human health. These compounds, traditionally derived from fish oil, are increasingly sourced directly from algae, providing a sustainable alternative for dietary supplements [133, 134]. In aquaculture, microalgae are already used as feedstock to enhance the nutritional profile of farmed fish and shellfish. Incorporating plankton-derived nutrients into human and animal diets holds promise for addressing malnutrition and improving global food security [135].

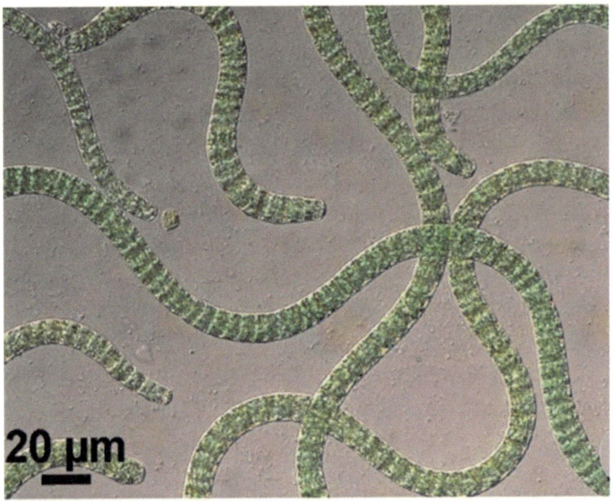

Figure 6.
Picture of Limnospira maxima *(formerly* Arthrospira maxima*) was kindly provided by the Culture Collection of Autotrophic Organisms (CCALA), http://ccala.butbn.cas.cz*

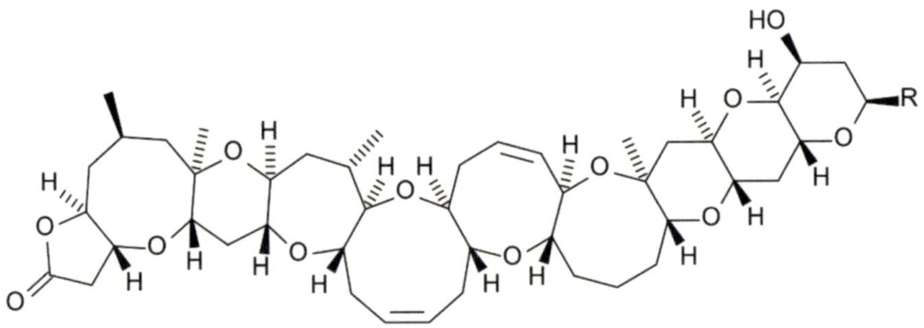

Figure 7.
Brevetoxin A, a group of neurotoxins isolated from the marine dinoflagellate Karenia brevis. *Image in Public Domain.*

5.11 Environmental applications

Plankton also offers solutions for environmental challenges. Their ability to sequester CO_2 has inspired efforts to cultivate phytoplankton on a large scale to mitigate climate change. Additionally, microalgae can be used in bioremediation to remove pollutants from wastewater. Genus like *Chlorella* and *Scenedesmus* are efficient in absorbing heavy metals and organic contaminants, helping to clean industrial effluents and reduce environmental pollution [136, 137].

5.12 Cosmetics and biotechnology

Plankton-derived compounds are increasingly incorporated into cosmetics and personal care products. For example, microalgal extracts are used in anti-aging creams, sunscreens, and moisturizers due to their high antioxidant content and ability to protect against UV radiation [138, 139]. In biotechnology, plankton enzymes are being explored for industrial applications. Algal enzymes, for instance, are used in the production of bioethanol, while others serve as catalysts in pharmaceutical synthesis. Furthermore, genetic engineering techniques applied to the plankton species have unlocked their potential for producing recombinant proteins and other valuable bioproducts [140].

5.13 Challenges and future prospects

Despite their potential, scaling up plankton-based technologies presents challenges. Large-scale cultivation requires optimization of growth conditions, efficient harvesting methods, and cost-effective processing techniques. Environmental concerns, such as the risk of harmful algal blooms from uncontrolled cultivation, must also be addressed [141]. Advances in synthetic biology and bioprocess engineering are paving the way for overcoming these challenges. Genetic modifications are enhancing the yield of desirable products, while innovations in photobioreactors are improving the efficiency of large-scale plankton cultivation [142]. Plankton's biotechnological applications span renewable energy, healthcare, nutrition, and environmental management, offering sustainable solutions to some of the world's pressing challenges. Continued research and development in this field hold immense promises for harnessing the full potential of these microscopic organisms, paving the way for a more sustainable and innovative future [143].

Acknowledgements

Leonel Pereira thanks to the Fundação para a Ciência e Tecnologia, I. P (FCT) in the framework of the Project UIDB/04004/2020 and DOI identifier 10.54499/UIDB/04004/2020 (https://doi.org/10.54499/UIDB/04004/2020 accessed on 30 July 2024).

The author acknowledges the assistance of Microsoft Copilot in language editing for certain sections of this chapter.

Conflict of interest

The author declares no conflict of interest.

Glossary

Bacterioplankton: Bacteria and archaea that inhabit the water column play a crucial role in the decomposition of organic matter and the recycling of nutrients.

Biogeochemical cycles: Cycles that involve the circulation of essential chemical elements between living beings and the non-living environment (land, water, and air).

Copepods (Copepoda): Small planktonic crustaceans that are an important food source for many fish and other marine organisms.

Cyanobacteria: A group of photosynthetic bacteria, also known as blue-green algae, that produce oxygen and can form blooms in nutrient-rich waters.

Diatoms (Bacillariophyceae): A group of photosynthetic unicellular algae with a silica cell wall, contributing significantly to oxygen production.

Dinoflagellates (Dinoflagellata): Single-celled algae with flagella that can be photosynthetic or heterotrophic and are known to cause red tides.

Euryhaline: Organisms that can tolerate a wide range of salinities, common in transitional environments such as estuaries.

Eutrophic: Nutrient-rich bodies of water promote abundant growth of phytoplankton and Cyanobacteria.

Holoplankton: Organisms that spend their entire lives in the form of plankton, such as diatoms and some species of jellyfish.

Meroplankton: Organisms that are planktonic only during one phase of their life cycle, such as starfish larvae.

Nutrient recycling: The process by which nutrients are reused in the ecosystem, often mediated by decomposer organisms.

Oligotrophic: Nutrient-poor bodies of water, often dominated by smaller plankton and diatoms.

Phytoplankton: Photosynthetic microorganisms considered "the grass of the sea", that produce oxygen and organic compounds through photosynthesis.

Plankton: Aquatic organisms that live suspended in the water column and have a limited ability to swim against currents.

Zooplankton: Heterotrophic organisms that feed on phytoplankton, other zooplankton or organic detritus, serving as primary consumers. Upwelling: The process of deep, nutrient-rich water rising to the ocean surface, boosting marine productivity.

Author details

Leonel Pereira
Marine Algae Lab, Department of Life Sciences, CFE—Centre for Functional Ecology: Science for People and Planet, Marine Resources, Conservation and Technology, University of Coimbra, Coimbra, Portugal

*Address all correspondence to: leonel.pereira@uc.pt

References

[1] Reynolds CS, Padisák J. Plankton, status and role of. In: Levin SA, editor. Encyclopedia of Biodiversity. 2nd ed. Waltham: Academic Press; 2013. pp. 24-38. DOI: 10.1016/B978-0-12-384719-5.00292-6

[2] McManus MA, Woodson CB. Plankton distribution and ocean dispersal. Journal of Experimental Biology. 2012;**215**(6):1008-1016. DOI: 10.1242/jeb.059014

[3] Jahan S. The role of phytoplanktons in the environment and in human life, a review. Basra Journal of Science. 2023;**41**:392-411. DOI: 10.29072/basjs.20230212

[4] Jeansson E, Bellerby RGJ, Skjelvan I, Frigstad H, Ólafsdóttir SR, Olafsson J. Fluxes of carbon and nutrients to the Iceland Sea surface layer and inferred primary productivity and stoichiometry. Biogeosciences. 2015;**12**(3):875-885. DOI: 10.5194/bg-12-875-2015

[5] El Gamal AA. Biological importance of marine algae. Saudi Pharmaceutical Journal. 2010;**18**(1):1-25. DOI: 10.1016/j.jsps.2009.12.001

[6] Fenchel T. Marine plankton food chains. Annual Review of Ecology, Evolution, and Systematics. 1988;**19**:19-38. DOI: 10.1146/annurev.es.19.110188.000315

[7] De La Rocha CL, Passow U. 8.4 - the biological pump. In: Holland HD, Turekian KK, editors. Treatise on Geochemistry. 2nd ed. Oxford: Elsevier; 2014. pp. 93-122. DOI: 10.1016/B978-0-08-095975-7.00604-5

[8] Lomartire S, Marques JC, Gonçalves AMM. The key role of zooplankton in ecosystem services: A perspective of interaction between zooplankton and fish recruitment. Ecological Indicators. 2021;**129**:107867. DOI: 10.1016/j.ecolind.2021.107867

[9] Castellani C. Plankton: A guide to their ecology and monitoring for water quality. Journal of Plankton Research. 2009;**32**(2):261-262. DOI: 10.1093/plankt/fbp102

[10] Miki T, Jacquet S. Complex interactions in the microbial world: Underexplored key links between viruses, bacteria and protozoan grazers in aquatic environments. Aquatic Microbial Ecology. 2008;**51**(2):195-208. DOI: 10.3354/ame01190

[11] Madsen EL. Microorganisms and their roles in fundamental biogeochemical cycles. Current Opinion in Biotechnology. 2011;**22**(3):456-464. DOI: 10.1016/j.copbio.2011.01.008

[12] Paerl HW, Justic D, Paerl RW. 4.13 - primary producers: Coastal phytoplankton ecology and trophic dynamics in the face of human and climatic pressures. In: Baird D, Elliott M, editors. Treatise on Estuarine and Coastal Science. 2nd ed. Oxford: Academic Press; 2024. pp. 348-373. DOI: 10.1016/B978-0-323-90798-9.00005-6

[13] Lawton G. Fungi ahoy! New Scientist. 2024;**261**(3477):37-39. DOI: 10.1016/S0262-4079(24)00274-4

[14] NOAA. COPEPOD - What Are Plankton and Why Are they Important? NOAA. Maryland: Silver Spring; 2024. Available from: https://www.st.nmfs.noaa.gov/copepod/about/what-n-why.html [Accessed: November 28, 2024]

[15] Sen K, Sen B, Wang G. Diversity, abundance, and ecological roles of planktonic fungi in marine environments. Journal of Fungi. 2022;**8**(5):491. DOI: 10.3390/jof8050491

[16] Xiao R, Wang Q, Zhang M, Pan W, Wang JJ. Plankton distribution patterns and the relationship with environmental gradients and hydrological connectivity of wetlands in the Yellow River Delta. Ecohydrology & Hydrobiology. 2020;**20**(4):584-596. DOI: 10.1016/j.ecohyd.2020.01.002

[17] Walks DJ, Cyr H. Movement of plankton through lake-stream systems. Freshwater Biology. 2004;**49**(6):745-759. DOI: 10.1111/j.1365-2427.2004.01220.x

[18] Kübar K, Agasild H, Virro T, Ott I. Vertical distribution of zooplankton in a strongly stratified hypertrophic lake. In: Ott I, Kõiv T, editors. Lake Verevi, Estonia — A Highly Stratified Hypertrophic Lake. Dordrecht: Springer Netherlands; 2005. pp. 151-162. DOI: 10.1007/1-4020-4363-5_13

[19] Visser AW, Stips A. Turbulence and zooplankton production: Insights from PROVESS. Journal of Sea Research. 2002;**47**(3):317-329. DOI: 10.1016/S1385-1101(02)00120-X

[20] Not F, Siano R, Kooistra WHCF, Simon N, Vaulot D, Probert I. Chapter one - diversity and ecology of eukaryotic marine phytoplankton. In: Piganeau G, editor. Advances in Botanical Research. Amsterdam: Academic Press; 2012. pp. 1-53. DOI: 10.1016/B978-0-12-391499-6.00001-3

[21] Carstensen J, Klais R, Cloern JE. Phytoplankton blooms in estuarine and coastal waters: Seasonal patterns and key species. Estuarine, Coastal and Shelf Science. 2015;**162**:98-109. DOI: 10.1016/j.ecss.2015.05.005

[22] Paczkowska J, Rowe OF, Figueroa D, Andersson A. Drivers of phytoplankton production and community structure in nutrient-poor estuaries receiving terrestrial organic inflow. Marine Environmental Research. 2019;**151**:104778. DOI: 10.1016/j.marenvres.2019.104778

[23] Vereshchaka A. Navigating the zooplankton realm: Oceans of diversity beneath the sea surface. Diversity. 2024;**16**(12):717. DOI: 10.3390/d16120717

[24] Ligorini V, Garrido M, Malet N, Simon L, Alonso L, Bastien R, et al. Response of phytoplankton communities to variation in salinity in a small Mediterranean coastal lagoon: Future management and foreseen climate change consequences. Water. 2023;**15**(18):3214. DOI: 10.3390/w15183214

[25] Muylaert K, Sabbe K, Vyverman W. Changes in phytoplankton diversity and community composition along the salinity gradient of the Schelde estuary (Belgium/the Netherlands). Estuarine, Coastal and Shelf Science. 2009;**82**(2):335-340. DOI: 10.1016/j.ecss.2009.01.024

[26] de Senerpont Domis LN, Elser JJ, Gsell AS, Huszar VLM, Ibelings BW, Jeppesen E, et al. Plankton dynamics under different climatic conditions in space and time. Freshwater Biology. 2013;**58**(3):463-482. DOI: 10.1111/fwb.12053

[27] NOAA. Aquatic Food Webs. National Oceanic and Atmosfere Administration. NOAA. Maryland: Silver Spring; 2024. Available from: https://www.noaa.gov/education/resource-collections/marine-life/aquatic-food-webs [Accessed: November 29, 2024]

[28] Falkowski PG. The ocean's invisible forest. Scientific American.

2002;**287**(2):54-61. DOI: 10.1038/scientificamerican0802-54

[29] Santos M, Moita MT, Oliveira PB, Amorim A. Phytoplankton communities in two wide-open bays in the Iberian upwelling system. Journal of Sea Research. 2021;**167**:101982. DOI: 10.1016/j.seares.2020.101982

[30] Wilken S, Soares M, Urrutia-Cordero P, Ratcovich J, Ekvall MK, Van Donk E, et al. Primary producers or consumers? Increasing phytoplankton bacterivory along a gradient of lake warming and browning. Limnology and Oceanography. 2018;**63**(S1):S142-SS55. DOI: 10.1002/lno.10728

[31] Gonzalez-Pestana A, Alfaro-Shigueto J, Mangel JC. A review of high trophic predator-prey relationships in the pelagic northern Humboldt system, with a focus on anchovetas. Fisheries Research. 2022;**253**:106386. DOI: 10.1016/j.fishres.2022.106386

[32] Botterell ZLR, Lindeque PK, Thompson RC, Beaumont NJ. An assessment of the ecosystem services of marine zooplankton and the key threats to their provision. Ecosystem Services. 2023;**63**:101542. DOI: 10.1016/j.ecoser.2023.101542

[33] Wang H, Liu Z, Li J, Lin B, Zhao Y, Zhang X, et al. Sinking fate and carbon export of zooplankton fecal pellets: Insights from time-series sediment trap observations in the northern South China Sea. Biogeosciences. 2023;**20**(24):5109-5123. DOI: 10.5194/bg-20-5109-2023

[34] D'alelio D, Montresor M, Mazzocchi MG, Margiotta F, Sarno D, Ribeira d'Alcalà M. Plankton food-webs: To what extent can they be simplified? Advances in Ocaenography and Limnology. 2016;**7**(1):67-92. Available from: https://www.pagepressjournals.org/aiol/article/view/5646 [Accessed: November 29, 2024]

[35] Mitra A, Castellani C, Gentleman WC, Jónasdóttir SH, Flynn KJ, Bode A, et al. Bridging the gap between marine biogeochemical and fisheries sciences; configuring the zooplankton link. Progress in Oceanography. 2014;**129**:176-199. DOI: 10.1016/j.pocean.2014.04.025

[36] Dai Y, Yang S, Zhao D, Hu C, Xu W, Anderson DM, et al. Coastal phytoplankton blooms expand and intensify in the 21st century. Nature. 2023;**615**(7951):280-284. DOI: 0.1038/s41586-023-05760-y

[37] Soule T, Garcia-Pichel F. Cyanobacteria. In: Schmidt TM, editor. Encyclopedia of Microbiology. 4th ed. Oxford: Academic Press; 2019. pp. 799-817. DOI: 10.1016/B978-0-12-809633-8.20886-8

[38] Basu S, Mackey KRM. Phytoplankton as key mediators of the biological carbon pump: Their responses to a changing climate. Sustainability. 2018;**10**(3):869. DOI: 10.3390/su10030869

[39] Naselli-Flores L, Padisák J. Ecosystem services provided by marine and freshwater phytoplankton. Hydrobiologia. 2023;**850**(12):2691-2706. DOI: 10.1007/s10750-022-04795-y

[40] Lobus NV, Kulikovskiy MS. The co-evolution aspects of the biogeochemical role of phytoplankton in aquatic ecosystems: A review. Biology. 2023;**12**(1):92. DOI: 10.3390/biology12010092

[41] Zafrilla B, Matarredona L, Bonete M-J, Zafrilla G, Esclapez J. Phytoplankton as CO_2 sinks: Redirecting the carbon cycle. Applied Sciences. 2024;**14**(19):8657. DOI: 10.3390/app14198657

[42] Baumas C, Bizic M. A focus on different types of organic matter particles and their significance in the open ocean carbon cycle. Progress in Oceanography. 2024;**224**:103233. DOI: 10.1016/j.pocean.2024.103233

[43] Zehr JP, Kudela RM. Nitrogen cycle of the open ocean: From genes to ecosystems. Annual Review of Marine Science. 2011;**3**:197-225. DOI: 10.1146/annurev-marine-120709-142819

[44] Steinberg DK, Saba GK. Chapter 26 - nitrogen consumption and metabolism in marine zooplankton. In: Capone DG, Bronk DA, Mulholland MR, Carpenter EJ, editors. Nitrogen in the Marine Environment. 2nd ed. San Diego: Academic Press; 2008. pp. 1135-1196. DOI: 10.1016/B978-0-12-372522-6.00026-8

[45] Pajares S, Ramos R. Processes and microorganisms involved in the marine nitrogen cycle: Knowledge and gaps. Frontiers in Marine Science. 2019;**6**:739. DOI: 10.3389/fmars.2019.00739

[46] Britannica. The Editors of Encyclopaedia: Phosphorus Cycle. Encyclopedia Britannica. Chicago: Britannica; 2020. Available from: https://www.britannica.com/science/phosphorus-cycle [Accessed: November 29, 2024]

[47] McMahon KD, Read EK. Microbial contributions to phosphorus cycling in Eutrophic lakes and wastewater. Annual Review of Microbiology. 2013;67:199-19219. DOI: 10.1146/annurev-micro-092412-155713

[48] Lønborg C, Müller M, Butler ECV, Jiang S, Ooi SK, Trinh DH, et al. Nutrient cycling in tropical and temperate coastal waters: Is latitude making a difference? Estuarine, Coastal and Shelf Science. 2021;**262**:107571. DOI: 10.1016/j.ecss.2021.107571

[49] Edwards TM, Puglis HJ, Kent DB, Durán JL, Bradshaw LM, Farag AM. Ammonia and aquatic ecosystems – A review of global sources, biogeochemical cycling, and effects on fish. Science of the Total Environment. 2024;**907**:167911. DOI: 10.1016/j.scitotenv.2023.167911

[50] Falkowski P. Ocean science: The power of plankton. Nature. 2012;**483**(7387):S17-S20. DOI: 10.1038/483S17a

[51] Ndah AB, Meunier CL, Kirstein IV, Göbel J, Rönn L, Boersma M. A systematic study of zooplankton-based indices of marine ecological change and water quality: Application to the European marine strategy framework directive (MSFD). Ecological Indicators. 2022;**135**:108587. DOI: 10.1016/j.ecolind.2022.108587

[52] Rao DB, Surendra T, Laxmi CNV, Meera KM, Gupta GVM, Kumar BSK. Effect of groundwater nutrients on coastal phytoplankton community composition in the bay of Bengal, India: An experimental study. Marine Pollution Bulletin. 2024;**209**:117016. DOI: 10.1016/j.marpolbul.2024.117016

[53] Yang X-e, Wu X, Hao H-l, He Z-l. Mechanisms and assessment of water eutrophication. Journal of Zhejiang University Science B. 2008;**9**(3):197-209. DOI: 10.1631/jzus.B0710626

[54] Steinberg DK, Landry MR. Zooplankton and the ocean carbon cycle. Annual Review of Marine Science. 2017;**9**:413-444. DOI: 10.1146/annurev-marine-010814-015924

[55] Edwards M, Richardson A. The impact of climate change on the phenology of the plankton community and trophic mismatch. Nature.

2004;**430**(7002):881-884. DOI: 10.1038/nature02808

[56] Negrete-García G, Luo JY, Petrik CM, Manizza M, Barton AD. Changes in Arctic ocean plankton community structure and trophic dynamics on seasonal to interannual timescales. Biogeosciences. 2024;**21**(22):4951-4973. DOI: 10.5194/bg-21-4951-2024

[57] Ruppert KM, Kline RJ, Rahman MS. Past, present, and future perspectives of environmental DNA (eDNA) metabarcoding: A systematic review in methods, monitoring, and applications of global eDNA. Global Ecology and Conservation. 2019;**17**:e00547. DOI: 10.1016/j.gecco.2019.e00547

[58] Lan J, Liu P, Hu X, Zhu S. Harmful algal blooms in eutrophic marine environments: Causes, monitoring, and treatment. Water. 2024;**16**(17):2525. DOI: 10.3390/w16172525

[59] Estes M, Anderson C, Appeltans W, Bax N, Bednaršek N, Canonico G, et al. Enhanced monitoring of life in the sea is a critical component of conservation management and sustainable economic growth. Marine Policy. 2021;**132**:104699. DOI: 10.1016/j.marpol.2021.104699

[60] Suthers I, Rissik D. The importance of plankton. In: Rissik D, Suthers I, editors. Plankton: A Guide to Their Ecology and Monitoring for Water Quality. Clayton, Victoria: CSIRO Publishing; 2009. Chapter 1,1. DOI: 10.1071/9780643097131. ISBN: 9780643097131

[61] Ali N, Khan MH, Ali M, Sidra AS, Khan A, et al. Insight into microplastics in the aquatic ecosystem: Properties, sources, threats and mitigation strategies. Science of the Total Environment. 2024;**913**:169489. DOI: 10.1016/j.scitotenv.2023.169489

[62] Padisák J, Naselli-Flores L. Phytoplankton in extreme environments: Importance and consequences of habitat permanency. Hydrobiologia. 2021;**848**(1):157-176. DOI: 10.1007/s10750-020-04353-4

[63] Roy S, Chattopadhyay J. Towards a resolution of 'the paradox of the plankton': A brief overview of the proposed mechanisms. Ecological Complexity. 2007;**4**(1):26-33. DOI: 10.1016/j.ecocom.2007.02.016

[64] Pan J. Marine phytoplankton diversity and ecology. In: Marine Biology: A Functional Approach to the Oceans and their Organisms. Boca Raton, Florida: CRC Press; 2021. pp. 43-71. DOI: 10.1201/9780429399244-5

[65] Feng J, Mazzei M, Di Gregorio S, Niccolini L, Vitiello V, Ye Y, et al. Marine copepods as a microbiome hotspot: Revealing their interactions and biotechnological applications. Water. 2023;**15**(24):4203. DOI: 10.3390/w15244203

[66] Bilbao J, Pavloudi C, Blanco-Rayón E, Franco J, Madariaga I, Seoane S. Phytoplankton community composition in relation to environmental variability in the Urdaibai estuary (SE Bay of Biscay): Microscopy and eDNA metabarcoding. Marine Environmental Research. 2023;**191**:106175. DOI: 10.1016/j.marenvres.2023.106175

[67] Hochfeld I, Hinners J. Phytoplankton adaptation to steady or changing environments affects marine ecosystem functioning. EGUsphere. 2024;**2024**:1-41. DOI: 10.5194/egusphere-2024-1246

[68] Brendonck L, De Meester L. Egg banks in freshwater

zooplankton: Evolutionary and ecological archives in the sediment. Hydrobiologia. 2003;**491**(1):65-84. DOI: 10.1023/A:1024454905119

[69] Kléparski L, Beaugrand G, Edwards M, Schmitt FG, Kirby RR, Breton E, et al. Morphological traits, niche-environment interaction and temporal changes in diatoms. Progress in Oceanography. 2022;**201**:102747. DOI: 0.1016/j.pocean.2022.102747

[70] Bahlburg D, Hüppe L, Böhrer T, Thorpe SE, Murphy EJ, Berger U, et al. Plasticity and seasonality of the vertical migration behaviour of Antarctic krill using acoustic data from fishing vessels. Royal Society Open Science. 2023;**10**(9):230520. DOI: 10.1098/rsos.230520

[71] Hall CAM, Lewandowska AM. Zooplankton dominance shift in response to climate-driven salinity change: A mesocosm study. Frontiers in Marine Science. 2022;**9**:861297. DOI: 10.3389/fmars.2022.861297

[72] Azani N, Ghaffar MA, Suhaimi H, Azra MN, Hassan MM, Jung LH, et al. The impacts of climate change on plankton as live food: A review. IOP Conference Series: Earth and Environmental Science. 2021;**869**(1):012005. DOI: 10.1088/1755-1315/869/1/012005

[73] Rasconi S, Winter K, Kainz MJ. Temperature increase and fluctuation induce phytoplankton biodiversity loss – Evidence from a multi-seasonal mesocosm experiment. Ecology and Evolution. 2017;7(9):2936-2946. DOI: 10.1002/ece3.2889

[74] Huber V, Adrian R, Gerten D. A matter of timing: Heat wave impact on crustacean zooplankton. Freshwater Biology. 2010;**55**(8):1769-1779. DOI: 10.1111/j.1365-2427.2010.02411.x

[75] Neun S, Hintz NH, Schröder M, Striebel M. Phytoplankton response to different light colors and fluctuation frequencies. Frontiers in Marine Science. 2022;**9**:824624. DOI: 10.3389/fmars.2022.824624

[76] Pereira L, Pardal M. Introductory chapter: General overview on oceanography. In: Leonel P, Miguel AP, editors. Oceanography. Rijeka: IntechOpen; 2024. p. Ch. 1. DOI: 10.5772/intechopen.113821

[77] Conroy JA, Steinberg DK, Thibodeau PS, Schofield O. Zooplankton diel vertical migration during Antarctic summer. Deep Sea Research Part I: Oceanographic Research Papers. 2020;**162**:103324. DOI: 10.1016/j.dsr.2020.103324

[78] Chakraborty P, Acharyya T, Raghunadh Babu PV, Bandyopadhyay D. Impact of salinity and pH on phytoplankton communities in a tropical freshwater system: An investigation with pigment analysis by HPLC. Journal of Environmental Monitoring. 2011;**13**(3):614-620. DOI: 10.1039/c0em00333f

[79] Komoroske LM, Jeffries KM, Connon RE, Dexter J, Hasenbein M, Verhille C, et al. Sublethal salinity stress contributes to habitat limitation in an endangered estuarine fish. Evolutionary Applications. 2016;**9**(8):963-981. DOI: 10.1111/eva.12385

[80] Bristow LA, Mohr W, Ahmerkamp S, Kuypers MMM. Nutrients that limit growth in the ocean. Current Biology. 2017;**27**(11):R474-R4R8. DOI: 10.1016/j.cub.2017.03.030

[81] Devlin M, Brodie J. Nutrients and eutrophication. In: Reichelt-Brushett A, editor. Marine Pollution – Monitoring, Management and

Mitigation. Cham: Springer Nature Switzerland; 2023. pp. 75-100. DOI: 10.1007/978-3-031-10127-4_4

[82] Wurtsbaugh WA, Paerl HW, Dodds WK. Nutrients, eutrophication and harmful algal blooms along the freshwater to marine continuum. WIREs Water. 2019;**6**(5):e1373. DOI: 10.1002/wat2.1373

[83] Karakuş O, Völker C, Iversen M, Hagen W, Hauck J. The role of zooplankton grazing and nutrient recycling for Global ocean biogeochemistry and phytoplankton phenology. Journal of Geophysical Research: Biogeosciences. 2022;**127**(10):e2022JG006798. DOI: 10.1029/2022JG006798

[84] Bişinicu E, Boicenco L, Pantea E, Timofte F, Lazăr L, Vlas O. Qualitative model of the causal interactions between phytoplankton, zooplankton, and environmental factors in the Romanian black sea. Phycology. 2024;**4**(1):168-189. DOI: 10.3390/phycology4010010

[85] Kath NJ. Functional traits determine biomass dynamics, coexistence and energetics in plankton food webs [thesis]. Germany: University of Potsdam; 2022. DOI: 10.25932/publishup-55123

[86] Chakraborty SK. Ocean ecosystem and its multidimensional eco-functionality and significance. In: Brinkmann R, editor. The Palgrave Handbook of Global Sustainability. Cham: Springer International Publishing; 2023. pp. 537-581. DOI: 10.1007/978-3-031-01949-4_37

[87] Sterner RW. Role of zooplankton in aquatic ecosystems. In: Likens GE, editor. Encyclopedia of Inland Waters. Oxford: Academic Press; 2009. pp. 678-688. DOI: 10.1016/B978-012370626-3.00153-8

[88] Declerck SAJ, de Senerpont Domis LN. Contribution of freshwater metazooplankton to aquatic ecosystem services: An overview. Hydrobiologia. 2023;**850**(12):2795-2810. DOI: 10.1007/s10750-022-05001-9

[89] Trathan PN, Hill SL. The importance of krill predation in the Southern Ocean. In: Siegel V, editor. Biology and Ecology of Antarctic Krill. Cham: Springer International Publishing; 2016. pp. 321-350. DOI: 10.1007/978-3-319-29279-3_9

[90] Glibert PM, Mitra A. From webs, loops, shunts, and pumps to microbial multitasking: Evolving concepts of marine microbial ecology, the mixoplankton paradigm, and implications for a future ocean. Limnology and Oceanography. 2022;**67**(3):585-597. DOI: 10.1002/lno.12018

[91] Flöder S, Burns CW. The influence of fluctuating light on diversity and species number of nutrient-limited phytoplankton. Journal of Phycology. 2005;**41**(5):950-956. DOI: 10.1111/j.1529-8817.2005.00124.x

[92] Wells ML, Karlson B, Wulff A, Kudela R, Trick C, Asnaghi V, et al. Future HAB science: Directions and challenges in a changing climate. Harmful Algae. 2020;**91**:101632. DOI: 10.1016/j.hal.2019.101632

[93] Giraldo C, Cresson P, MacKenzie K, Fontaine V, Loots C, Delegrange A, et al. Insights into planktonic food-web dynamics through the lens of size and season. Scientific Reports. 2024;**14**(1):1684. DOI: 10.1038/s41598-024-52256-4

[94] Otero J, Álvarez-Salgado XA, Bode A. Phytoplankton diversity effect on ecosystem functioning in a coastal upwelling system. Frontiers in Marine

Science. 2020;7:592255. DOI: 10.3389/fmars.2020.592255

[95] Kazamia E, Helliwell KE, Purton S, Smith AG. How mutualisms arise in phytoplankton communities: Building eco-evolutionary principles for aquatic microbes. Ecology Letters. 2016;**19**(7):810-822. DOI: 10.1111/ele.12615

[96] Frainer A, McKie BG, Amundsen P-A, Knudsen R, Lafferty KD. Parasitism and the biodiversity-functioning relationship. Trends in Ecology & Evolution. 2018;**33**(4):260-268. DOI: 10.1016/j.tree.2018.01.011

[97] Sheppard CRC, Davy SK, Pilling GM. 984 Symbiotic Interactions. The Biology of Coral Reefs. Oxford: Oxford University Press; 2009. pp. 98-129. DOI: 10.1093/acprof:oso/9780198566359.003.0004

[98] Mutalipassi M, Riccio G, Mazzella V, Galasso C, Somma E, Chiarore A, et al. Symbioses of cyanobacteria in marine environments: Ecological insights and biotechnological perspectives. Marine Drugs. 2021;**19**(4):227. DOI: 10.3390/md19040227

[99] Lesser MP, Stat M, Gates RD. The endosymbiotic dinoflagellates (*Symbiodinium* sp.) of corals are parasites and mutualists. Coral Reefs. 2013;**32**(3):603-611. DOI: 10.1007/s00338-013-1051-z

[100] Li S-Q, Xiao Y, Huang H-L, Luo Z, Sun T-R, Gao H-Y, et al. Experimental evidence that nitrogen-fixing cyanobacterium *Trichodesmium* spp. supplies new nitrogen source to marine phytoplankton. Frontiers in Marine Science. 2024;**11**:1388725. DOI: 10.3389/fmars.2024.1388725

[101] Andersen K, Berge T, Gonçalves R, Hartvig M, Heuschele J, Hylander S, et al. Characteristic sizes of life in the oceans, from bacteria to whales. Annual Review of Marine Science. 2015;**8**:1-25. DOI: 10.1146/annurev-marine-122414-034144

[102] Sime-Ngando T. Phytoplankton *Chytridiomycosis*: Fungal parasites of phytoplankton and their imprints on the food web dynamics. Frontiers in Microbiology. 2012;**3**:361. DOI: 10.3389/fmicb.2012.00361

[103] Rasconi S, Grami B, Niquil N, Jobard M, Sime-Ngando T. Parasitic chytrids sustain zooplankton growth during inedible algal bloom. Frontiers in Microbiology. 2014;**5**:229. DOI: 10.3389/fmicb.2014.00229

[104] Yoshiyama K, Mellard Jarad P, Litchman E, Klausmeier CA. Phytoplankton competition for nutrients and light in a stratified water column. The American Naturalist. 2009;**174**(2):190-203. DOI: 10.1086/600113

[105] Mei Z-P, Finkel ZV, Irwin AJ. Light and nutrient availability affect the size-scaling of growth in phytoplankton. Journal of Theoretical Biology. 2009;**259**(3):582-588. DOI: 10.1016/j.jtbi.2009.04.018

[106] Schlüter L, Møhlenberg F, Havskum H, Larsen S. The use of phytoplankton pigments for identifying and quantifying phytoplankton groups in coastal areas: Testing the influence of light and nutrients on pigment/chlorophyll a ratios. Marine Ecology-progress Series - Marine Ecology Progress Series. 2000;**192**:49-63. DOI: 10.3354/meps192049

[107] Gilpin LC, Davidson K, Roberts E. The influence of changes in nitrogen: Silicon ratios on diatom growth dynamics. Journal of Sea Research.

2004;**51**(1):21-35. DOI: 10.1016/j.seares.2003.05.005

[108] Poulson KL, Sieg RD, Prince EK, Kubanek J. Allelopathic compounds of a red tide dinoflagellate have species-specific and context-dependent impacts on phytoplankton. Marine Ecology Progress Series. 2010;**416**:69-78. DOI: 10.3354/meps08788

[109] Elena L. Competition and coexistence of phytoplankton under fluctuating light: Experiments with two cyanobacteria. Aquatic Microbial Ecology. 2003;**31**(3):241-248. DOI: 10.3354/ame031241

[110] Johnson KH, Vogt KA, Clark HJ, Schmitz OJ, Vogt DJ. Biodiversity and the productivity and stability of ecosystems. Trends in Ecology & Evolution. 1996;**11**(9):372-377. DOI: 10.1016/0169-5347(96)10040-9

[111] Silva A. Standing at water edges: Ecohydrological interactions between coastal groundwater discharge and intertidal community dynamics. Marine Environmental Research. 2024;**202**:106762. DOI: 10.1016/j.marenvres.2024.106762

[112] Falkowski PG. The role of phytoplankton photosynthesis in global biogeochemical cycles. Photosynthesis Research. 1994;**39**(3):235-258. DOI: 10.1007/BF00014586

[113] Nowicki M, DeVries T, Siegel DA. Quantifying the carbon export and sequestration pathways of the ocean's biological carbon pump. Global Biogeochemical Cycles. 2022;**36**(3):e2021GB007083. DOI: 10.1029/2021GB007083

[114] Blaser M, Conrad R. Stable carbon isotope fractionation as tracer of carbon cycling in anoxic soil ecosystems. Current Opinion in Biotechnology. 2016;**41**:122-129. DOI: 10.1016/j.copbio.2016.07.001

[115] Ziveri P, Gray WR, Anglada-Ortiz G, Manno C, Grelaud M, Incarbona A, et al. Pelagic calcium carbonate production and shallow dissolution in the North Pacific Ocean. Nature Communications. 2023;**14**(1):805. DOI: 10.1038/s41467-023-36177-w

[116] Riebesell U, Zondervan I. Reduced calcification of marine plankton in response to increased atmospheric CO_2. Nature. 2000;**407**:2-5. DOI: 10.1038/35030078

[117] Archibald KM, Siegel DA, Doney SC. Modeling the impact of zooplankton diel vertical migration on the carbon export flux of the biological pump. Global Biogeochemical Cycles. 2019;**33**(2):181-199. DOI: 10.1029/2018GB005983

[118] Cavicchioli R, Ripple WJ, Timmis KN, Azam F, Bakken LR, Baylis M, et al. Scientists' warning to humanity: Microorganisms and climate change. Nature Reviews Microbiology. 2019;**17**(9):569-586. DOI: 10.1038/s41579-019-0222-5

[119] Heinrichs ME, Mori C, Dlugosch L. Complex interactions between aquatic organisms and their chemical environment elucidated from different perspectives. In: Jungblut S, Liebich V, Bode-Dalby M, editors. YOUMARES 9 - the Oceans: Our Research, our Future: Proceedings of the 2018 Conference for YOUng MArine RESearcher in Oldenburg, Germany. Cham: Springer International Publishing; 2020. pp. 279-297. DOI: 10.1007/978-3-030-20389-4_15

[120] Lizotte M, Levasseur M, Law CS, Walker CF, Safi KA, Marriner A, et al.

Dimethylsulfoniopropionate (DMSP) and dimethyl sulfide (DMS) cycling across contrasting biological hotspots of the New Zealand subtropical front. Ocean Science. 2017;**13**(6):961-982. DOI: 10.5194/os-13-961-2017

[121] Beardall J, Stojkovic S, Larsen S. Living in a high CO_2 world: Impacts of global climate change on marine phytoplankton. Plant Ecology & Diversity. 2009;**2**(2):191-205. DOI: 10.1080/17550870903271363

[122] López-Abbate MC. Microzooplankton communities in a changing ocean: A risk assessment. Diversity. 2021;**13**(2):82. DOI: 0.3390/d13020082

[123] Fabry VJ, Seibel BA, Feely RA, Orr JC. Impacts of ocean acidification on marine fauna and ecosystem processes. ICES Journal of Marine Science. 2008;**65**(3):414-432. DOI: 10.1093/icesjms/fsn048

[124] Visser PM, Verspagen JMH, Sandrini G, Stal LJ, Matthijs HCP, Davis TW, et al. How rising CO_2 and global warming may stimulate harmful cyanobacterial blooms. Harmful Algae. 2016;**54**:145-159. DOI: 10.1016/j.hal.2015.12.006

[125] Nellemann C, Corcoran E, Duarte C, Valdes L, Young C, Fonseca L, et al. Blue Carbon - the Role of Healthy Oceans in Binding Carbon. Arendal, Norway: GRID; 2009. Available from: https://www.grida.no/publications/145 [Accessed: November 29, 2024]

[126] Landrigan PJ, Stegeman JJ, Fleming LE, et al. Human health and ocean pollution. Annals of Global Health. 2020;**86**(1):151. DOI: 10.5334/aogh.2831

[127] Pereira L, Cotas J, Valado A. Antioxidants from microalgae and their potential impact on human well-being. Exploration of Drug Science. 2024;**2**(3):292-321. DOI: 10.37349/eds.2024.00048

[128] Tomar S, Agarwal S, Singh H, Kumar R, Qureshi KA, Jaremko M, et al. Microalgae: A promising source for biofuel production. Biocatalysis and Agricultural Biotechnology. 2023;**53**:102877. DOI: 10.1016/j.bcab.2023.102877

[129] Dolganyuk V, Belova D, Babich O, Prosekov A, Ivanova S, Katserov D, et al. Microalgae: A promising source of valuable bioproducts. Biomolecules. 2020;**10**(8):1153. DOI: 10.3390/biom10081153

[130] Fernandes T, Cordeiro N. Microalgae as sustainable biofactories to produce high-value lipids: Biodiversity, exploitation, and biotechnological applications. Marine Drugs. 2021;**19**(10):573. DOI: 10.3390/md19100573

[131] Eze CN, Onyejiaka CK, Ihim SA, Ayoka TO, Aduba CC, Jk N, et al. Bioactive compounds by microalgae and potentials for the management of some human disease conditions. AIMS Microbiology. 2023;**9**(1):55-74. DOI: 10.3934/microbiol.2023004

[132] Kini S, Divyashree M, Mani MK, Mamatha BS. Chapter 12 - algae and cyanobacteria as a source of novel bioactive compounds for biomedical applications. In: Singh PK, Kumar A, Singh VK, Shrivastava AK, editors. Advances in Cyanobacterial. Biology: Academic Press; 2020. pp. 173-194. DOI: 10.1016/B978-0-12-819311-2.00012-7

[133] Cho K, Heo J, Han J, Hong HD, Jeon H, Hwang H-J, et al. Industrial

applications of dinoflagellate phycotoxins based on their modes of action: A review. Toxins. 2020;**12**(12):805. DOI: 10.3390/toxins12120805

[134] La Barre S, Bates SS. Marine biominerals with a biotechnological future. Blue Biotechnology. In: Blue Biotechnology: Production and Use of Marine Molecules. Hoboken, New Jersey, USA: Wiley-VCH Verlag; 2018. pp. 855-912. Chapter 27. DOI: 10.1002/9783527801718.ch27

[135] Gower LB. Biomimetic model systems for investigating the amorphous precursor pathway and its role in biomineralization. Chemical Reviews. 2008;**108**(11):4551-4627. DOI: 10.1021/cr800443h

[136] Barta DG, Coman V, Vodnar DC. Microalgae as sources of omega-3 polyunsaturated fatty acids: Biotechnological aspects. Algal Research. 2021;**58**:102410. DOI: 10.1016/j.algal.2021.102410

[137] Wells ML, Potin P, Craigie JS, Raven JA, Merchant SS, Helliwell KE, et al. Algae as nutritional and functional food sources: Revisiting our understanding. Journal of Applied Phycology. 2017;**29**(2):949-982. DOI: 10.1007/s10811-016-0974-5

[138] Kusmayadi A, Leong YK, Yen H-W, Huang C-Y, Chang J-S. Microalgae as sustainable food and feed sources for animals and humans – Biotechnological and environmental aspects. Chemosphere. 2021;**271**:129800. DOI: 10.1016/j.chemosphere.2021.129800

[139] Al-Thani RF, Yasseen BT. Cyanoremediation of Polluted Seawater in the Arabian Gulf: Risks and Benefits to Human Health. Processes. 2024;**12**:2733. DOI: 10.3390/pr12122733

[140] Devi A, Saran C, Saratale GD, Saratale RG, Ferreira LFR, Mulla SI, et al. Sustainable algal industrial wastewater treatment. Microbes Based Approaches for the Management of Hazardous Contaminants. 2024:190-205. DOI: 10.1002/9781119851158.ch12

[141] Jahan A, Ahmad IZ, Fatima N, Ansari VA, Akhtar J. Algal bioactive compounds in the cosmeceutical industry: A review. Phycologia. 2017;**56**(4):410-422. DOI: 10.2216/15.58.1

[142] Rotter A, Varamogianni-Mamatsi D, Zvonar Pobirk A, Gosenca Matjaž M, Cueto M, Díaz-Marrero AR, et al. Marine cosmetics and the blue bioeconomy: From sourcing to success stories. iScience. 2024;**27**(12):111339. DOI: 10.1016/j.isci.2024.111339

[143] Rotter A, Barbier M, Bertoni F, Bones AM, Cancela ML, Carlsson J, et al. The essentials of marine biotechnology. Frontiers in Marine Science. 2021;**8**:629629. DOI: 10.3389/fmars.2021.629629

Chapter 2

Complex Interactions in the Lake Kinneret Ecosystem

Moshe Gophen

Abstract

Published models indicate that phytoplankton density, and consequently water quality is dependent on grazing capacity by herbivore zooplankton (small and large Cladocera and Rotifera). Moreover, the top-down eco-force, cascading effect, produced by zooplanktivore fishes, Cyclopoid predator zooplankton, or both, attributes the principal pressure. Nevertheless, complex interactions within the ecosystem have also indicated an impact induced by other factors such as nutrients (bottom-up eco-force), which affect the major food resource of herbivore zooplankton, and grazeable algae (Chlorophyta, Diatoms) density. Temperature significantly affects zooplankton density as well as lake water residence time and Water Level. Two methods of statistical analyzes were utilized aimed at the evaluation of a multivariate comprised ecosystem: Principal Component Analyzes (PCA) and its illustrated plot (Biplot). Results conclusively indicate that zooplanktivore fishes (Sardines) and temperature are the Principal Components. Seclude of isolated single factor as a unique impacting parameter on zooplankton density, either predator Cyclopoida or Sardine fishes is therefore misleading.

Keywords: Kinneret, predation, nutrients, Sardine, Cyclopoida, herbivore zooplankton, temperature

1. Introduction

Statistical analyzes of Principal Component Analysis (PCA) and Graphical Representation of a PCA that combine both the scores and loading into a single plot (Biplot) was carried out using the software of STATA 17.0-Standard Edition, Statistics and Data Science, Copyright 1985–2021 StataCorp LLC, 4905 Lakeway Drive, 800-STATA-PC, Stata license: Single-user perpetual, Serial number: 401706315938, Licensed to Moshe Gophen, Migal. The life cycle of Copepoda within the zooplankton community comprised 10–12 stages, of which the first five are nauplii and taxonomic definitions were not applied in routine sample analysis. Within the next 10–11 copepodite stages the last four and five stages, as well as adults, were routinely defined as Cyclopoida. Nauplii and 1–3 of copepodite stages are herbivores, while 4–5 copepodite stages and adult copepods (termed as Cyclopoida) are predators. Consequently, the "Copepoda" variable includes all life cycle stages, and "Cyclopoida"—the predator stages. The fish, namely "Sardine" variable, includes two endemic bleak species, the most common fish in Lake Kinneret creating the heaviest predation pressure on

Parameters	Units
Annual means, Rotifera, Cladocera, Copepoda density	$G(ww)/m^2$
Predator Cyclopoida, herbivore Copepeoda	No/L
Small (1–3 neonates) and large (Adult) Cladocera	No/L
Small/large Cladocera ratio	No/L
Diaphanosoma sp., *Bosmina* spp., *Ceridaphnia* spp.	No/L
Herbivores zooplankton production	$(gC/m^2/month)$
Herbivores zooplankton grazing	$(gC/m^2/month)$
Chlorophyta, Diatoms, *Peridinium* density	$g(ww)/\ m^2$
TN, TP, TN/TP mass ratio	ppm; Ton
Water Level (annual mean) (higher value = lower level)	mbsl
Primary production	$(gC/m^2/day)$
Annual mean epilimnion temperature	°C
Monthly mean residence time	Year values
Sardine annual harvest	Ton/Year
Annual means, Rotifera, Cladocera, Copepoda density	$G(ww)/m^2$

Table 1.
Environmental variables and their units.

zooplankton: Common Hebrew names are Lavnoon Kinneret and Lavnoon lisner and both are "Sardine" variable; *Mirogrex terraesanctae* and *Acanthobrama lissneri*. The Lake Kinneret limnological data sources were applied by the Lake Kinneret Data Base and annual reports of the Kinneret Limnological Laboratory, IOLR, 1969–2001. Information about the Sardine (Bleaks) annual harvest was applied by the annual reports of the Department of Fishery, Agriculture Ministry, Lake Kinneret Fishery Unit 1969–2001. Variable list and their units are given in **Table 1**.

2. Results

Eight different variable combinations of the environmental parameters (**Table 1**) were analyzed by Biplot and PCA and are presented in **Figures 1–8**.

Results given in **Table 2** indicate a high level of explained variance by the two coordinates of the PCA analysis, as illustrated by the components in Biplots illustrations. Moreover, the higher the number of variables, the lower the value of the Total Explained Variance (V). Moreover, the assumption of top-down eco-force, as grazing capacity by herbivore zooplankton (small and large Cladocera and Rotifera) is just one variable (predator cyclopoids or Sardine fishes) attribution, the explained variance of the grazer's density is high. Nevertheless, the complexity of interactions within the ecosystem justifies the involvement of surplus other factors which has an impact on the major food resource of herbivore zooplankton, grazeable algae (Chlorophyta, Diatoms) density. Consequently, several environmental factors (**Table 2**) were involved in PCA/Biplot analyzes (**Figures 1–8; Table 1**).

DOI: http://dx.doi.org/10.5772/intechopen.1006143

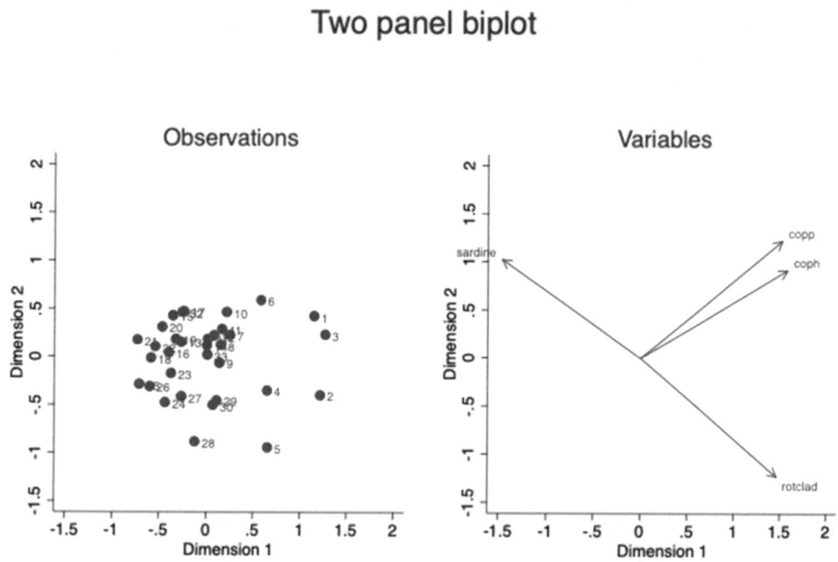

Principal components (eigenvectors)

Variable	Comp1	Comp2	Comp3	Comp4
rotclad	--------	-0.5559	0.4995	0.4539
coph	0.5228	---------	0.4749	-0.5740
copp	0.5042	0.5524	----------	0.5487
sardine	-0.4868	0.4628	0.6208	----------

Figure 1.
Densities (No/L) of total Rotifera plus Cladocera; herbivore Copepoda, predator Copepoda (Cyclopoida), and Sardines (ton, annual harvest).

Results of eight PCA analyzes and Biplot illustrations of paired variables that were randomly selected are presented in **Figures 1–8**. The positive and negative (−) Eigenvector values are given.

The negative (inverse) correlations were sorted into five classes (A) of correlated variables.

A: Lake loads of Nutrients, Lake Water Level, Lake Residence Time, and Epilimnetic Temperature correlated with Zooplankton communities.

B: Lake Nutrient Loads correlated with Phytoplankton.

C: Zooplankton correlated with Phytoplankton.

D: Correlative relations in-between Zooplankton communities.

E: Zooplankton and Phytoplankton correlated with Fish.

A summary of Eigenvectors that were calculated for each paired correlation sorted by class is shown in **Table 3**.

Results given in **Table 3** indicate a high level of data variability, as expressed by the 38–71% range of the SDs from the mean. The high level of variability resulting

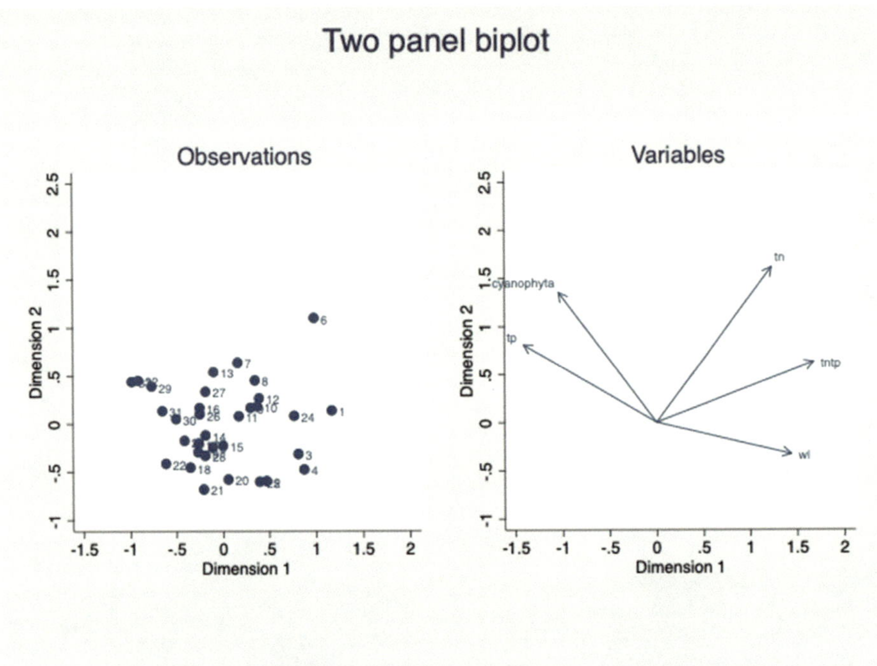

Principal components (eigenvectors)

Variable	Comp1	Comp2	Comp3	Comp4	Comp5
tn	---------	0.6819	0.3463	-0.0446	-0.5033
tp	-0.4603	---------	0.5944	0.2912	0.4822
tntp	0.5439	0.2653	----------	-0.3182	0.7132
wl	0.4666	-0.1393	-0.0443	---------	0.0742
cyanophyta	-0.3390	0.5724	-0.7078	0.2377	---------

Figure 2.
Lake load (ton) of total nitrogen (TN), total phosphorus (TP), and TN/TP mass ratio.

from these tested paired correlated parameters comprise the structured multivariate communities creating the Kinneret complex interaction system. Seclude of isolated single factor as a unique impacting parameter on zooplankton density, for example, either predator Cyclopoida or Sardine fishes is therefore misleading. Nevertheless, an indication that the Sardine factor, among others including Cyclopoida, is a dominant or principal component is justified.

3. Discussion

The study of ecological interactions is commonly disputed between two major concepts: (1) on an individual basis, specifically between one independent environmental parameter in relation to one or several dependent factors [1–3]. (2) Modeling construction which is supplied by experimental algorithms [4–8]. The first is limited between one influence and one or more influencers' parameters. Models are evaluated by experimental algorithms that are changeable and might be therefore flexible.

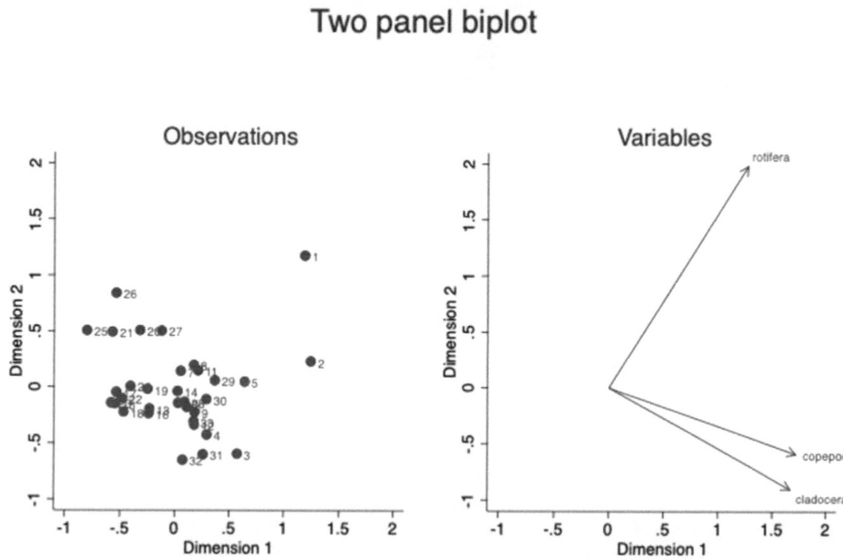

Principal components (eigenvectors)

Variable	Comp1	Comp2	Comp3
copepoda	--------	-0.2609	-0.7276
cladocera	0.6154	---------	0.6797
rotifera	0.4677	0.8790	--------

Figure 3.
Biomass (g(ww)/m²) of Copepoda, Cladocera and Rotifera.

Moreover, an old discussion among limnologists was dedicated to the issue defined as each lake is different as phrased by the popular proverb "But in My Lake" [9]. A relevant case study was carried out in Lake Kinneret during the late 1990s to early 2000s. The lake suffered from overwhelmed zooplankton predation pressure by the most common fish in the lake. A recommendation was submitted to the governmental authorities to subsidize the removal of unwanted sardine fishes aimed at reducing predation pressure from herbivore Cladocera to enhance algal grazing capacity for the improvement of water quality. This recommendation was accepted and several thousand tons of noncommercial Sardines were removed. Nevertheless, besides the significant recovery of the Cladocera community, algal biomass did not decline. The reason was indicated; the reduction of Phosphorus inputs was not considered. The factor that enhanced the outbreak of Cyanobacteria in Lake Kinneret (1994) is not solely nitrogen deficiency [1, 2] or Phosphorus sufficiency [3] and algal biomass reduction is not only reduction of their grazing pressure and the density of herbivore Cladocera is not the only dependent of predator cyclopoids. The ecological instructive lesson given by those few case studies is that in the management of an ecosystem comprised of complex interaction, a multivariate evaluation is required and the PCA

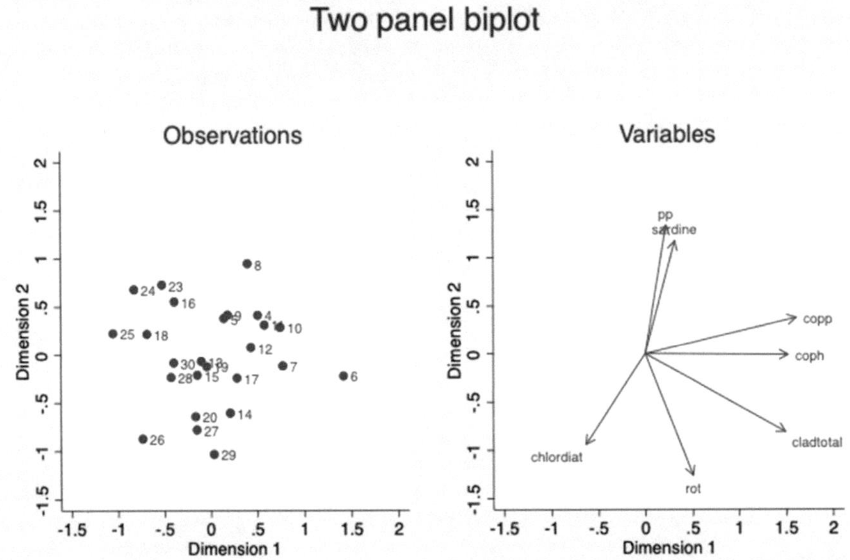

Principal components (eigenvectors)

Variable	Comp1	Comp2	Comp3	Comp4	Comp5	Comp6	Comp7
sardine	---------	-0.4621	0.5333	0.1362	0.6859	0.0125	-0.0210
coph	0.5401	---------	0.1480	0.2217	-0.2272	-0.7625	0.0651
copp	0.5710	-0.1484	--------	-0.1702	-0.1677	0.4614	0.6175
rot	0.1778	0.4978	-0.4474	--------	0.5977	-0.0205	0.2402
cladtotal	0.5300	0.3209	0.1528	-0.0187	----------	0.3470	-0.6869
chlordiat	-0.2260	0.3711	0.5367	0.6071	-0.2485	--------	0.2153
pp	0.0790	-0.5253	-0.4260	0.6552	-0.1752	0.1955	---------

Figure 4.
Sardine (ton, annual harvest), densities (No/L) of herbivore Copepoda, predator Copepoda (Cyclopoida), total Cladocera; biomass (g(ww)/m^2) of Chloropyta and Diatoms, and primary production (gC/m^2/day).

and Biplot analysis presented here is the answer. The utilization of the PCA and Biplot method avoidance of manipulated data management was carried out.

PCA analysis represents the directions of the loaded data that explain the maximum amount of variance. Those directions are presented as arrowed lines that capture most of the loaded data. The higher the variance included by a line, the larger the dispersion of the data points along it and the more information it has. This results in a better visibility of the differences between the observations. Moreover, the located position of the arrowed lines also indicates the correlation within variables: Positively correlated – the two tested variables are correlated, increase, or decrease together. Negatively or inversely correlated – among the two tested variables when one increases, the other decreases, and vice versa. The Eigenvector of the covariance matrix is a PCA component. The Eigenvalues are the coefficients attached to the Eigenvector which give the amount of variance carried in each Principal Component. The Eigenvector represents

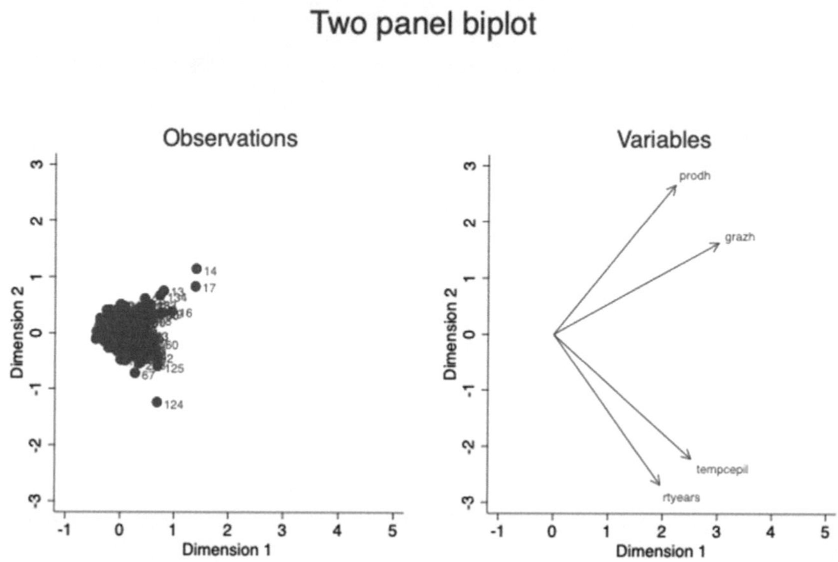

Principal components (eigenvectors)

Variable	Comp1	Comp2	Comp3	Comp4
tempcepil	--------	0.4746	-0.5637	0.4422
rtyears	0.3980	--------	0.6876	-0.1993
grazh	0.6135	-0.3490	---------	-0.6607
prodh	0.4513	-0.5692	0.3796	---------

Figure 5.
Epilimnion temperature; lake residence time; herbivore grazing, and herbivore production ($gC/m^2/month$).

the direction of the axes where there is the most variance information. The Eigenvector value represents the best fit of the line direction, with maximum variance indicated. The Eigenvalue is the number representing the data spread on the line, the Eigenvector. The larger the Eigenvalue the higher the impact on the tested object. Data given in **Table 3** consequently indicate a higher impact of Sardine fishes on herbivore zooplankton density than the impact of Cyclopoida predation.

Complex interactions research within a lake ecosystem is therefore an appropriate stimulator for comparative scientific bridging between versatile variables supported by multivariate evaluation. Positive correlations between paired parameters from different disciplines (nutrient, phytoplankton, zooplankton, fishes, physical trait etc.) that were indicated between variables within **Figures 1–8** imply another factor such as temperature, mutual predator, or mutual food resource which has a similar impact on the two others. Nevertheless, negative (inverse) correlations between paired parameters from different disciplines (nutrient, phytoplankton, zooplankton, fishes, physical traits, etc.) point to contrasting relations implying environmental significance. The ecosystem's complex interactions include interfingering through

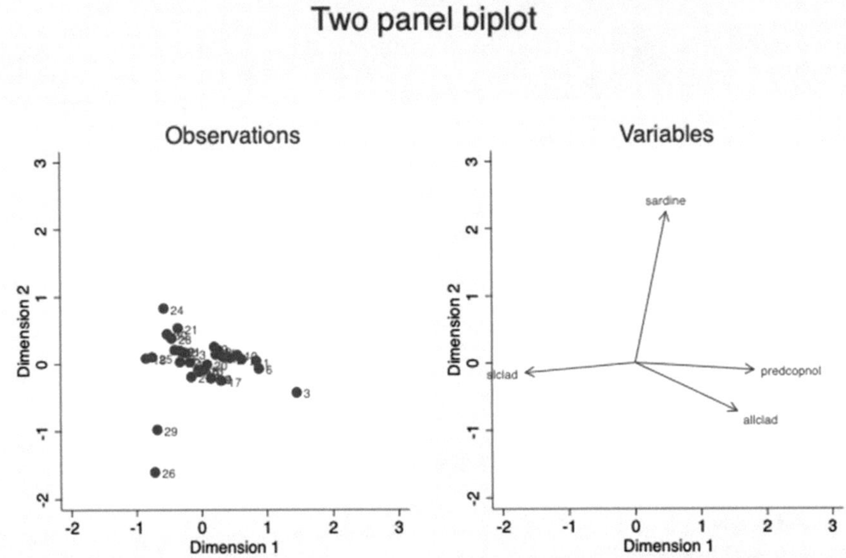

Principal components (eigenvectors)

Variable	Comp1	Comp2	Comp3	Comp4
slclad	---------	-0.0615	0.6546	0.4979
allclad	0.5263	---------	0.7047	-0.3662
predcopnol	0.6142	-0.0458	---------	0.7847
sardine	0.1614	0.9496	0.2645	---------

Figure 6.
Densities (No/L): Small/large Cladocera ratio, all Cladocera, predator Copepoda (Cyclopoida), and Sardine (ton, annual harvest).

many interlocking processes which partly or completely overlap and therefore confound multifarious structures. The following inverse relations in paired variables with respect to the Figure illustrations were indicated.

PCA analysis of the fish (Sardine) predation predator Cyclopoida impact on the density of herbivore Zooplankton (**Figure 1**) resulted in two inverse relations:

Sardine vs. densities of herbivore zooplankton (Rotifera plus Cladocera) and Herbivore zooplankton vs. Herbivore Copepods. It is suggested that zooplankton densities are dependent on Sardine biomass. Herbivore Copepoda is in the early life cycle stages of predator Cyclopoida. Therefore, the potential impact of other than fish predation on zooplankton is possible [4, 5, 7, 10–15]. A principal issue in the complex interaction of freshwater ecosystems is who the dependents of phytoplankton biomass are. Nutrients (bottom-up) or grazers (top-down). During the 1990s, a significant change in the phytoplankton community structure occurred when Cyanobacteria replaced *Peridinium* as the dominant phytoplankton component. Consequently, the impact of total nitrogen and total phosphorus, as well as lake Water Level fluctuations on Cyanobacteria biomass, was

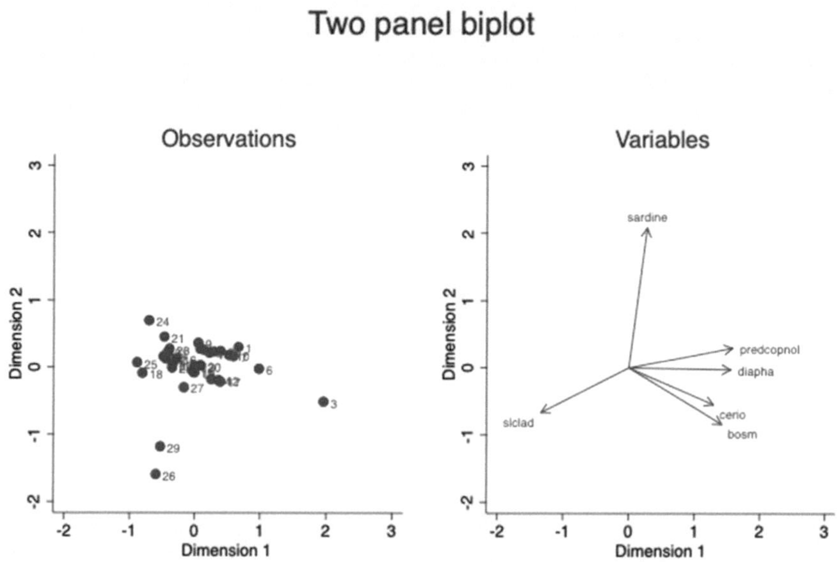

Principal components (eigenvectors)

Variable	Comp1	Comp2	Comp3	Comp4	Comp5	Comp6
slclad	---------	-0.2747	0.5498	0.4795	0.2251	0.4117
diapha	0.4799	----------	-0.0106	0.3435	0.7427	-0.3161
bosm	0.4382	-0.3460	--------	0.4736	-0.6096	-0.2459
cerio	0.3987	-0.2258	0.6244	---------	0.0763	0.0978
predcopnol	0.4893	0.1234	-0.2701	0.1218	---------	0.8103
sardine	0.0847	0.8593	0.4504	0.1680	-0.1394	---------

Figure 7.
Densities (No/L): Small/large Cladocera ratio, Diaphansoma sp., Bosmina spp., Ceriodaphnia spp., Predator Copepoda (Cyclopoida), and Sardine (ton, annual harvest).

analyzed (**Figure 2**). *Peridinium* is not edible to herbivore zooplankton and most of the Cyanobacteria as well. Nevertheless, Chlorophyta and Diatoms are edible and favored by herbivore zooplankton. Consequently, a PCA-Biplot analysis was carried out to evaluate the relations between Cyanobacteria and TP, TN, and Water Levels. Inverse relations of TN vs. TP, Cyanobacteria Biomass vs. TN, Water Level vs. TP, Cyanobacteria biomass vs. TN/TP mass ratio, and Tn/TP Mass ratio vs. Water Level. Earlier studies confirmed that the appearance and fluctuated biomass in Lake Kinneret are dependent on TN and TP lake loads regime and probably not affected by zooplankton grazing [1–3]. Nevertheless, Water Level is indirectly involved. The lower the level altitude (**Figure 2**): the higher the numbers below sea level, lake location) is the lower TP load as affected by changes in climate conditions (precipitation and river discharge decline) [8].

Among several prime issues that are predicted to be involved within the Kinneret ecosystem complexed interactions, the optional of intra-community predation

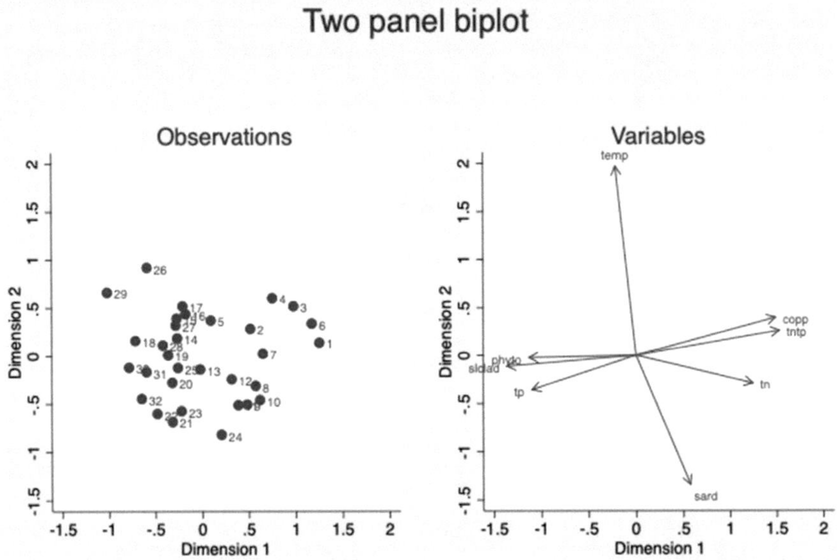

Principal components (eigenvectors)

Variable	Comp1	Comp2	Comp3	Comp4	Comp5	Comp6	Comp7	Comp8
temp	---------	0.7954	0.2671	0.2700	-0.0105	0.4658	-0.0399	-0.0094
phyto	-0.3440	----------	0.3907	0.5636	-0.1564	-0.6007	-0.1531	-0.0502
tn	0.3782.	-0.1182.	----------	0.5462	0.4066	0.0951	-0.2783	-0.4405
tp	-0.3348	-0.1444	-0.5381	----------	-0.2281	0.2553	0.2353	0.4218
tntp	0.4643	0.1050	0.1204	0.1358	---------	-0.2346	0.0439	0.7310
sard	0.1743	-0.5421	0.5974	0.2098	-0.1258	---------.	0.1134	0.0478
slclad	-0.4145	-0.0455	0.1199	0.0039	0.6872	0.0091	---------	-0.1447
copp	0.4520	0.1597	-0.0436	0.1471	-0.3477	-0.2212	0.7134	----------

Figure 8.
Epilimnion temperature, total phytoplankton biomass (g(ww)/m²); lake load (ton) of TN, TP, TN/TP mass ratio; Sardine (ton, annual harvest); densities (No/L) of small/large Cladocera, and Predator Copepoda (Cyclopoida).

was analyzed through biomass (g(ww)/m^2) densities of Copepoda, Cladocera, and Rotifera (**Figure 3**). Copepoda was considered to include predator life cycle stages (Cyclopoida) and Cladocera and Rotifera as herbivores [16–23]. PCA-Biplot analysis results have indicated a significant impact of the biomass of Copepoda on the biomass of Rotifera and, to a lesser extent, on Cladocera. It probably resulted from an indirect effect of top-down eco-force pressure. Several studies have documented the selectivity of large body size prey by visual predators by fishes (Sardine and young stages Tilapias) in Lake Kinneret. Large body zooplankters are preyed more efficiently [7, 11, 16–23]. The decline of preferable consumption of large body size predator Cyclopoida was therefore accompanied by enhancement of smaller body size rotifers and, to a lesser extent, young neonates Cladocera [7].

A step forward was carried out as a multivariate internal correlation between phytoplankton, zooplankton, and fish variables (**Figure 4**):

PCA/Biplot figure number	Number of variables	Explained variance by component (Coordinate)1	Explained variance by component (Coordinate)2	Total explained variance
1	4	0.6414	0.1866	0.8280
2	5	0.5606	0.1952	0.7558
3	3	0.5620	0.2704	0.8324
4	7	0.3750	0.2450	0.6200
5	4	0.4707	0.3878	0.8585
6	4	0.6015	0.2511	0.8526
7	6	0.5877	0.1855	0.7732
8	8	0.4735	0.1517	0.6252

Table 2.
Number of variables, values of explained variance by coordinate (Component) 1 and 2, and the total explained variance resulted by the PCA/Biplot.

Class	Mean	SD	%	Minimum	Maximum	n
A	0.3032	0.1399	46	0.0443	0.4603	8
B	0.1997	0.0924	46	0.1182	0.3390	5
C	0.3345	0.2370	71	0.1752	0.6869	4
D	0.3347	0.2137	61	0.0458	0.7275	13
E	0.4492	0.1722	38	0.2212	0.6869	8

Table 3.
The class (A–E) averaged Eigenvalues of the negatively paired correlated (inversely) variables (see **Figures 1–8***) are summarized as mean class values (mean), SD's, and its % (%), of the mean, maximum, minimum, and a number of parameters (n) are given.*

The following inverse correlations between ecosystem variables were indicated: Between the biomass density of edible algal groups (Chlorophyta plus Diatoms) and Sardine, interpreted that if Sardine biomass was increased, resulted in an intensification of zooplankton predation created, grazing pressure was reduced and consequently algal biomass enhancement; An inverse correlation between the Primary Production (PP) of total Phytoplankton and the density of edible algal and Herbivore life stages of Copepoda (nauplii and copepodite stages) was indicated. Until the mid-1990s the dominant phytoplankter was the un-edible bloom-forming *Peridinium* and edible algal groups (Chlorophyta and Diatoms) were therefore suppressed. The PP enhancement is due to the *Peridinium* biomass while edible algal biomass was diminished.

Further inverse correlations were indicated between Sardine biomass vs. Herbivore Copepoda density and the density of Herbivore vs. Predator Copepoda; The density of Rotifera vs. Predator Copepoda; The density of Total Cladocera vs. Herbivore Copepoda.

The density of Total Cladocera vs. Predator Copepoda; The density of Total Cladocera vs. biomass density of edible phytoplankters (Chlorophyta plus Diatoms); Total Cladocera and Primary Production; Herbivore Copepoda density vs. biomass density of edible algal groups (Chlorophyta and Diatoms); Density of Total Cladocera vs. the biomass density of edible algal groups (Chlorophyta and Diatoms); and finally, between the Biomass density of edible algal groups (Chlorophyta and Diatoms) vs. Primary Production reflecting the positive correlation between total biomass and PP of *Peridinium* dominated Phytoplankton assemblages.

Inverse correlation between lake hydraulic residence time (RT) (years) and epilimnetic temperature vs. metabolic active capacities (grazing and production) of herbivore and predator Cyclopoida zooplankton were identified (**Figure 5**): Residence Time vs. herbivore grazing capacity (gC/m^2/month); Residence Time vs. predator Cyclopoida production.

Herbivores grazing capacity vs. predator Production. It is concluded that these inverse correlations indicate that RT prolongation was accompanied by temperature increase and temperature elevation enhanced metabolic activity of zooplankton (herbivores and predators). Temperature increase is likely a result of RT prolongation as a consequence of seasonality [14, 16, 17, 24–28].

A critical issue was tested through the correlation between fish (Sardine) biomass and the density of the total number of Cladocera which resulted in significant inverse relations (**Figure 6**). A low statistical probability of the correlation between the density of the total number of Cladocera and Small/Large Cladocera ratio vs. predator Cyclopoida was also indicated. The correlation between fish biomass and Cladocera density was significantly high. Consequently, the intensive fish predation pressure on Cladocera which is more efficient than that of Cyclopoida pressure on Cladocera was confirmed. The known cascading top-down fish predation pressure accompanied by lower pressure of Cyclopoida on Cladocera is confirmed.

The required sensible step forward is aimed at the test of indicative Cladocera prey selectivity by predator fish or Inverse correlation Cyclopoida as well as an in-between crustacean (*Bosmina* spp., *Ceriodaphnia* spp., Diaphanosoma sp., and predator Cyclopoida) community densities (**Figure 7**). Significant inverse correlations of the following were indicated: Sardine vs. *Diaphanosoma* sp., Sardine vs. *Bosmina* spp., *Bosmina* spp. vs. Diaphanosoma sp., *Ceripodaphnia* spp. vs. Diaphanosoma sp., and Cyclopoida vs. *Bosmina* spp. It has to be considered that food item selection by visual attackers zooplantivore fishes is dependent on body size and highly affected by skillful escapability. Therefore, Diaphanosoma and Cyclopoida are better escapers and less vulnerable than *Bosmina* and *Ceriodaphnia*. Moreover, the inverse correlation that was indicated between Cyclopoida and *Bosmina* might be a result of an indirect predation effect where Sardin visual attack predation selectively preyed more *Bosmina* and *Ceriodaphnia* while *Diaphanosoma* and Cyclopoida were therefore enhanced. A microscopical survey confirmed negligible residual fragments of *Bosmina* in the Cyclopoida gut content [16].

The complexity of the Kinneret ecosystem comprises verified variables of nutrients (TN, TP, TN/TP mass ratio) zooplankton densities, (Cladocera, Small/Large body size ratio of Cladocera, Cyclopoida), phytoplankton biomass distribution, and epilimnetic temperature (**Figure 8**) are presented as the following inverse correlations: Temperature vs. Phytoplankton, TP, Small/Large; Phytoplankton vs. epilimnetic temperature; Phytoplankton vs. TN, TP, Sardine, Small/Large cladoceran life cycle neonates; TN vs. TP, and Cyclopoida density; TN/TP mass ratio vs. Phytoplankton, TP, Sardine, and Cyclopoida density; Sardine vs. Phytoplankton, TN/TP, and Cyclopoida density; Small/Large of cladoceran life cycle neonates vs. Phytoplankton and TN; and finally Cyclopoida density vs. TN, Small/Large cladoceran life cycle stages ratio.

Those indicated inverse correlations reflect the temporal regular dynamics sequence pattern in the Kinneret ecosystem prior to the outbreak of Cyanobacteria (1994).

Starting during the second half of the winter after heavy external inputs of nitrogen (TN) during the winter which induce the formation of the dominant *Peridinium* bloom creating a high biomass of total phytoplankton which transfer Phosphorus from bottom sediments through the germination of dormant Cysts into free-swimming vegetative cells. Nevertheless, due to the second part of the winter q continuation of a spawning season of the Sardine fishes, their zooplankton predation is not yet enhanced while for the Cladocera it is the optimal season for reproduction resulting in high densities of young neonates and high-value of Small/Large body size ratio [5, 17, 29].

4. Conclusion

The basic assumption attached to the utilized statistical method of PCA/Biplot was: positive relation between two environmental variables within the complexity of the Kinneret ecosystem are mutual dependents of one or more factors. Whereas, if the correlation is inverse, these two (or more) variables might have an impact of one on the other, and if the correlation is positive, there is an involvement of another variable on both. Moreover, the higher the significance values, the stronger the impact of one on other factors. Paired variables were randomly selected within eight groups of PCA/Biplot analysis. The bilateral impact between two inversed variables maybe done through a third (or more) variable indirectly. An inverse correlation between the Cladocera *Bosmina* and predator Cyclopoida is created through the selectivity of the predator by Sardine. Seasonal or temporal elevation of epilimnetic temperature and/or climate condition changes clearly affect the metabolic activity and population density of Cladocera, and also nutrient inputs (TN, TP) into lake Kinneret and consequently replacement of dominant algal bloom of *Peridinium* by Cyanophyta.

Four model types aimed at comprehensive (physical and biological traits) structuring of the ecological complexity of the Kinneret ecosystem were published: (1) The Carbon flow pattern [5]; (2) A numerical simulation of the role of zooplankton in C, N, and P cycling [6]; (3) ECOPATH [4]; and (4) Intraguild predation dynamics based on a coupled Hydrodynamic-Ecological model [7]. Three of the models, 1st–3rd, postited zooplankton predation by Sardine as a solid cardinal status within the Kinneret ecosystem. The 4th model type attributed the major impact of Cyclopoida on zooplankton consumption. The model presented in this paper eliminated the usage of metabolic variables and ecological value algorithms, and the only usage of statistical distribution was evaluated. The level or value ranges of eco-physiological traits were replaced by statistical distribution through the PCA/Biplot method. Conclusively, the present study is in agreement with conclusions presented by 1–3 models and disagrees with the conclusion of model No. 4. Nevertheless, this conclusive summary does not confound the side effect of other variables on the distribution of zooplankton communities in Lake Kinneret.

Acknowledgements

Sincere appreciation is given to staff members and field data assistance of Kinneret Limnological Laboratory, Israel Oceanographic and Limnological Research Co. Ltd., Haifa, Lake Kinneret Data Base, and The Kinneret and Drainage Authority for availabilities of Data records, Migal-scientific research Institute in the Galilee, and Mekorot Water Co. Ltd.

Funding

This research received no external funding.

Author contributions

The author directed data evaluation and presentation design, computerization, and the preparation of the original and final draft version.

Conflicts of interest

The author declares no conflict of interest.

Data availability statement

The data presented in this study are available on request from the corresponding author.

Author details

Moshe Gophen
MIGAL - Scientific Research Institute in Galilee, Kiryat Shmone, Israel

*Address all correspondence to: gophen@migal.org.il

References

[1] Gophen M, Smith VH, Nishri A, Threlkeld ST. Nitrogen deficiency, phosphorus sufficiency, and the invasion of lake Kinneret, Israel, by N2-fixing cyanobacterium *Aphanizomenon ovalisporum*. Aquatic Sciences. 1999;**1**:1-14

[2] Smith VH. Low nitrogen to phosphorus ratios favor dominance by blue-green algae in lake phytoplankton. Science. 1983;**221**:669-671

[3] Hadas O, Pinkas R, Malinsky-Rushansky N, Nishri A, Kaplan A, Rimmer A, et al. Appearance and establishment of diazotrophic cyanobacteria in lake Kinneret, Israel. Freshwater Biology. 2012;**57**:1214-1227. DOI: 10.1111/j.1365-2427.2012.02792.x

[4] Walline P, Pizanty S, Gophen M, Berman T. The Ecosystem of Lake Kinneret, Israel. ICES.C.M. 1990/L:39. 1990. pp. 1-8

[5] Serruya C, Gophen M, Pollingher U. Lake Kinneret: Carbon flow patterns and ecosystem management. Archiv für Hydrobiologie. 1980;**88**(3):265-302

[6] Bruce LC, Hamilton D, Imberger J, Gal G, Gophen M, Zohary T, et al. A numerical simulation of the role of zooplankton in C, N, and P, cycling in lake Kinneret, Israel. Ecological Modelling. 2006;**193**(2006):412-436

[7] Makler V, Hipsey MR, Zohary T, Carmel Y, Gal G. Intraguild predation dynamics in a lake ecosystem based on a coupled hydrodynamic-ecological model: The example of lake Kinneret (Israel); MDPI. Biology. 2017, 2017;**6**(2):22

[8] Gal G, Skerianer M, Tanasova N. Fluctuations in water level and the dynamics of zooplankton: A data driven modeling approach. Freshwater Biolgy. 2013;**58**(4):800-816. DOI: 10.1111/fwb.12087

[9] Carpenter SR, editor. Complex Interactions in Lake Communities. New York Berlin Heildelberg London Tokyo: Springer-Verlag; 1988. 283 p

[10] Gophen M. The impact of zooplankton status on the management of lake Kinneret (Israel). Hydrobiologia. 1984;**113**:249-258

[11] Gophen M. Effect of fish predation on size class distribution of Cladocerans in lake Kinneret. Verhandlungen des Internationalen Verein Limnologie. 1985;**22**:3104-3108

[12] Gophen M. Fisheries management in lake Kinneret (Israel). In: Proc. Annual Meeting North American Lake Management Society; 13-16 November 1985; Lake Geneva, Wisconsin, USA. Wisconsin: NALMS; 1986. pp. 327-332

[13] Gophen M, Scharf A. Food and feeding habits of Mirogrex fingerlings in lake Kinneret (Israel). Hydrobiologia. 1981;**78**(1):3-9

[14] Gophen M, Serruya S. Fish predation effects on plankton community structure in large lakes: Lake Kinneret. In: Tilzer M, Serruya C, editors. Ecological Structure and Function in Large Lakes. Berlin, Heidelberg, New-York, London, Paris, Tokyo, Hong Kong: Springer-Verlag; 1990. pp. 645-660

[15] Easton J, Gophen M. Trophic relations between zooplankton and bleaks (*Acanthobrama* spp.) in lake Kinneret (Israel). Verhandlungen des Internationalen Verein Limnologie. 2002;**28**:1258-1261

[16] Gophen M. The metabolism of adult *Mesocyclops leuckarti* populations in lake Kinneret (Israel) during 1969-1978. Verhandlungen des Internationalen Verein Limnologie. 1981a;**21**:1568-1572

[17] Gophen M. Metabolic activity of herbivorous zooplankton in lake Kinneret (Israel) during 1972-1977. Journal of Plankton Research. 1981b;**3**(1):15-24

[18] Gophen M. Population dynamics of cyclopoid copepods in lake Kinneret (Israel). Open Journal of Modern Hydrology. 2016;**6**:212-221

[19] Gophen M, Threlkeld ST. An experimental study of zooplankton consumption by the lake Kinneret Sardine. Archiv für Hydrobiologie. 1989;**115**:91-95

[20] Drenner RW, Vinyard GL, Gophen M, McComas SR. Feeding behaviour of the cichlid Sarotherodon galilaeus selective predation on lake Kinneret Zooplankton. Hydrobiologia. 1982;**87**(1):17-20

[21] Drenner RW, Hambright JD, Vinyard GL, Gophen M, Pollingher U. Experimental study of size-selective phytoplankton grazing by a filter-feeding cichlid and the cichlid's effects on plankton community structure. Limnology and Oceanography. 1987;**32**(5):1138-1144

[22] Vinyard GL, Drenner RW, Gophen M, Pollingher U, Winkleman DL, Hambright KD. An experimental study of the plankton community impacts of two filter-feeding cichlids, the Galilee Saint Peter's fish (*Sarotherodon galilaeus*) and blue tilapia (*Tilapia aurea*). Canadian Journal of Fisheries and Aquatic Sciences. 1988;**45**(4):685-690

[23] Landau R, Gophen M, Walline P. Larval *Mirogrex terraesanctae* (Cyprinidae) of lake Kinneret (Israel): Growth rate, plankton selectivities, consumption rates and interactions with rotifers. Hydrobiologia. 1988;**169**:91-106

[24] Gophen M. Food and feeding habits of *Mesocyclops leuckarti* (Claus) in lake Kinneret (Israel). Freshwater Biology. 1977;7:513-518

[25] Gophen M. Artemia nauplii as a food source for cyclopoids: Extrapolation of experimental measurements to the metabolic activities of copepods in lake Kinneret, Israel. In: Persoone G, Sorgeloos P, Roels O, Jaspers E, editors. The Brine Shrimp Artemia. Vol. 3. Wetteren, Belgium: Universe Press; 1980. pp. 67-76

[26] Gophen M, Azoulay B. The trophic status of zooplankton communities in lake Kinneret (Israel). Verhandlungen des Internationalen Verein Limnologie. 2002;**28**:836-839

[27] Gophen M, Serruya S, Spataru P. Zooplankton community changes in lake Kinneret (Israel), during 1969-1985. In: Biro P, Talling JF, editors. Trophic Relationships in Inland Waters. Hydrobiologia. Vol. 191. 1990. pp. 39-46

[28] Gophen M, Serruya S. Effects of fish and nutrients on long-term changes of plankton communities in lake Kinneret (Israel). In: Schroder JH, Bauer J, Schartl M, editors. Trends in Ichthyology, an International Perspectives. Munich: Blackwell Scientific Publications; 1993. pp. 371-380

[29] Gophen M. Long-term changes of plankton-communities in lake Kinneret, Israel. Asian Fisheries Science. 1992;**5**:291-302

Chapter 3

Long-Term Shifts in Phytoplankton Community Composition and Abundance in Lake Champlain: A 50-Year Comparison

Timothy B. Mihuc, Kayleen Snyder, Zachary Cutter and Shelly Bouyea

Abstract

In this study, we examined long-term patterns in phytoplankton community composition and abundance in Lake Champlain from 1970 to 2021. Seven study sites were sampled across three time periods (1970, 2003–2005, and 2017–2021). Phytoplanktons were identified to the lowest possible taxonomic level, typically genus. Our results illustrate differences in phytoplankton community composition across the three time periods, particularly between 1970 and the two recent time periods. In 1970, diatoms and Chromista were dominant in phytoplankton assemblages while in latter time periods: 2003–2005 and 2017–2021. Diatoms remained abundant while cyanobacteria increased in abundance. Within each phytoplankton functional group, there were changes in the dominant phytoplankton genera from 1970 to present. Notably, *Dolichospermum* (formerly *Anabaena*) was the dominant Cyanobacteria in 1970, while *Microcystis* abundance increased in 2003–2005 and 2017–2021. The diatom community also shifted with increasing abundance of *Fragilaria* in 2017–2021. Phytoplankton community composition was similar across study sites within each study period suggesting the observed community patterns are lake wide. Shifts in the phytoplankton assemblage and genera from 1970 to the present were likely a result of climate change-induced water temperature increases and invasive species impacts, favoring taxa adapted to warmer conditions in recent years.

Keywords: Lake Champlain, phytoplankton, community dynamics, long-term patterns, climate change

1. Introduction

Lake Champlain is among the largest freshwater lakes in the United States, with a diverse phytoplankton assemblage [1]. It is located between New York and Vermont, USA, and extends into Québec, Canada [2]. The drainage basin is 21,326km^2 and consists predominantly of forests and agriculture [3]. This temperate, stratified dimictic lake is commonly divided into five main basins with varying depths and trophic status, resulting in a range of habitats for biotic communities [1]. Several studies have collected long-term data on phytoplankton community composition and abundance in Lake Champlain. Some of the earliest known collections of phytoplankton occurred in 1929 [4] and 1970, which revealed diatoms as one of the dominant taxonomic groups in the phytoplankton assemblage [1]. Comparisons between these early time periods showed a long-term shift from oligotrophic to mesotrophic-eutrophic phytoplankton species in sections of the lake [1]. Diatoms (Bacillariophyceae) were the predominant phytoplankton group in Lake Champlain during 1991–1992, along with an abundance of perennial cryptophyte flagellates while Cyanobacteria and green algae were present at low abundance [5].

The Laurentian Great Lakes' long-term monitoring has shown similar trends in phytoplankton assemblages over time [6]. In the 1970s, diatoms were the dominant phytoplankton group in most basins of Lake Erie, accompanied by a high presence of Chromista (Chrysophyceae, Cryptophyceae, and Dinoflagellata) [7]. Comparably, Lake Superior had a phytoplankton community comprised mostly protozoa and diatoms in 1973 [8]. In 1983, Lake Huron, Michigan, and Erie were sampled from spring to fall and found diatoms continued to have the highest biovolumes, while Cyanobacteria had the greatest overall densities [9]. Diatoms, Cyanobacteria, and Chlorophyta remained the dominant phytoplankton groups in Lake Erie during late summer of 2003–2005 [10]. More recently, in an 11-year study of the Great Lakes, spring diatom abundance was noted to decline in Lake Huron and Lake Michigan, whereas cyanobacteria cell densities increased in Lake Superior and Erie [6].

Climate change and other human-induced environmental changes are considered among the greatest threats to large freshwater lake ecosystems [11, 12]. Summer surface water temperatures have risen for several large lakes worldwide [13]. In the Lake Champlain Basin, the average summer air and surface water temperatures have increased over time [14, 15]. Higher surface temperatures in Lake Champlain during mid-summer months could lead to stronger thermal stratification and result in less mixing in the long term [15]. As temperatures continue to rise, Lake Champlain could experience fewer lake freeze-over events during winter, along with more intense but irregular storm occurrences [14, 16]. Changes in the lake's thermal structure could drive shifts in phytoplankton community composition and abundance, with important implications for water quality and the aquatic food web [16, 17]. Additionally, Lake Champlain is particularly susceptible to a nutrient runoff because of its large watershed, with a watershed-to-surface area ratio of 18:1 [1, 2].

Aquatic invasive species represent another major threat to lake ecosystems and can impact food web dynamics and reduce the resiliency and biodiversity of an ecosystem [18–20]. In Lake Champlain, there are >50 invasive species including fish, plants, mollusks, crustaceans, and pathogens [21, 22]. Aquatic invasive species have dispersed through a variety of transportation methods like ballast water, canals attached to boats and trailers, aquarium trade, and baitfish [22]. One of the most impactful invasive species that enter Lake Champlain is the zebra mussel (*Dreissena polymorpha*). Zebra mussels were first reported in Lake Champlain in 1993 and quickly populated throughout the lake [23]. These efficient filter feeders have been noted to directly reduce phytoplankton biovolume, rotifer abundance, and chlorophyll a and increase water clarity [18, 24, 25]. Mihuc and Recknagel [26] noted that zebra mussels preyed upon slow-moving rotifers and likely graze phytoplankton. Likewise, nitrogen and phosphorus concentrations may be reduced by zebra mussels, which indirectly impacts phytoplankton biomass [25]. More recently, spiny water flea (*Bythotrephes*

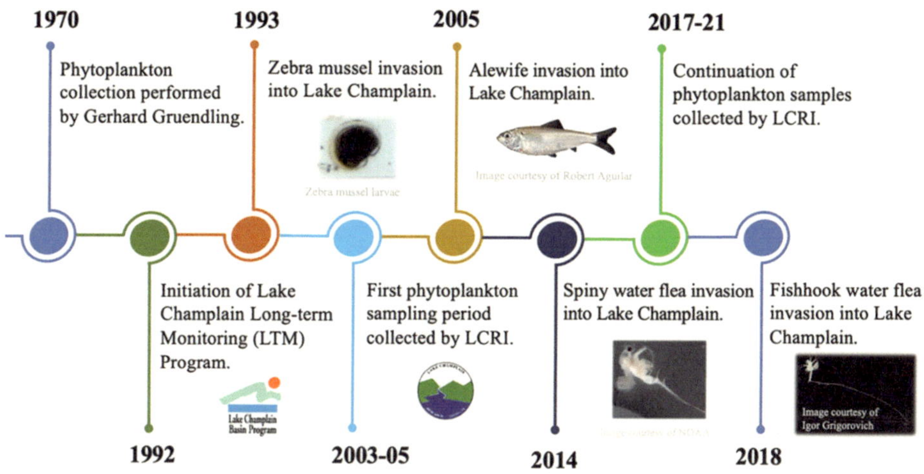

Figure 1.
Timeline of sampling periods and major events in Lake Champlain.

longimanus) and fishhook water flea (*Cercopagis pengoi*) invaded Lake Champlain in 2014 and 2018, respectively [26, 27]. Spiny water flea have been noted for reducing cyclopoid copepods and cladocerans species [28]. Comparably, in Lake Mendota, following the invasion of zebra mussels and spiny water flea Cyanobacteria dominance occurred earlier in the growing season [29]. These two invasive species are considered generalist feeders, which outcompete native zooplankton predators like *Mysis diluviana* (Arthropoda) [30].

The objective of this study was to examine long-term patterns in phytoplankton abundance and community composition in Lake Champlain. Historical phytoplankton data from 1970 [1] were compared to recent data (2003–2005 and 2017–2021). Study sites included five sites in the Main Lake and one in both St. Albans Bay and Malletts Bay. Overall, long-term monitoring is useful for understanding patterns and revealing potential drivers of change in the ecosystem over time [31]. Summary of Lake Champlain major environmental events (**Figure 1**) illustrates how monitoring can address patterns over multiple decades and environmental issues.

2. Methods

2.1 Study sites

The seven study sites were selected in Lake Champlain based on historical data collected in 1970 (**Figure 2**) [1]. Each site varies by depth and trophic status (**Table 1**). Phytoplankton samples were collected at each site from May to September for all sampling periods (1970, 2003–2005, and 2017–2021).

2.2 Field and laboratory procedures

Phytoplankton samples were collected in 1970 using a Van-Dorn bottle sample, at a vertical depth of 5 m [1]. The samples were filtered through a #20 silk bolting cloth (63.5 μm micron mesh) and placed into preservation vials with 95% ethyl alcohol for

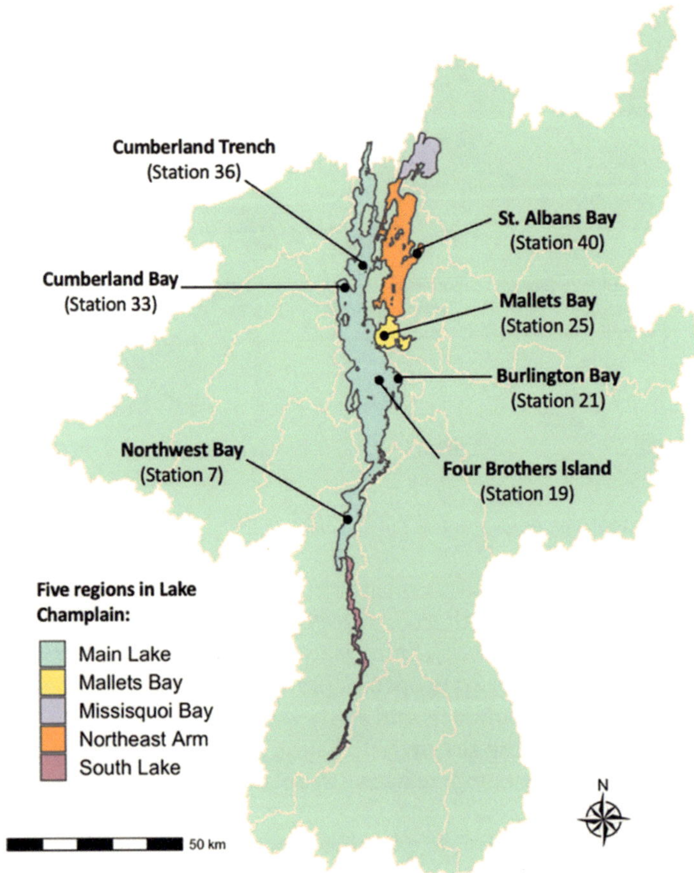

Figure 2.
Map of the seven monitoring sites in the Lake Champlain Basin.

Site #	Site name	Depth (m)	Trophic status	Latitude	Longitude
7	Northwest Bay	50	Oligo-mesotrophic	44°7′33.60"N	73°24′46.20"W
19	Four Brothers Island	100	Oligo-mesotrophic	44°28′15.60"N	73°17′57.00"W
21	Burlington Bay	15	Oligo-mesotrophic	44°28′29.40"N	73°13′54.00"W
25	Malletts Bay	32	Mesotrophic	44°34′55.20"N	73°16′52.20"W
33	Cumberland Bay	11	Oligo-mesotrophic	44°42′4.20"N	73°25′5.40"W
36	Cumberland Trench	50	Oligo-mesotrophic	44°45′22.20"N	73°21′18.00"W
40	St. Albans Bay	7	Eutrophic	44°47′7.20"N	73° 9′43.80"W

Trophic status based on 1977 trophic status report from EPA [32].

Table 1.
Physical characteristics of the seven study sites in Lake Champlain.

preservation. About 45 phytoplankton samples were collected for the seven study sites during 1970. The Lake Champlain Research Institute collected the 2003–2005 and 2017–2021 samples using a 63.5-μm micron Wisconsin plankton net with a 13 cm opening [33]. Samples were collected at twice the Secchi depth during the recent time periods and preserved with Lugol's iodine solution. The Secchi depth allows for measurement of water clarity and serves as an indicator for the phytoplankton biomass and suspended materials are in the water column [34]. The number of samples collected during 2003–2005 and 2017–2021 varied, with a total count of 60 and 310, respectively. Samples were stored in a dark room until laboratory analysis.

In the laboratory, phytoplankton samples were thoroughly mixed, and subsamples were pipetted from a centrifuge tube into a Sedgewick-Rafter counting cell. The Sedgewick-Rafter cell holds a 1 mL capacity with 100 counting cells. Phytoplankton were identified to the lowest feasible taxonomic level, using an inverted microscope. Each sample was counted until a minimum of 10 fields or 100 of the most abundant phytoplankter had been assessed, or up to three 1 mL aliquots were examined [33]. All samples in our study required a single aliquot. Phytoplankton samples were stored for 5 years.

2.3 Data analysis

Phytoplankton data were analyzed using R programming language version 4.2.2. Seventy-four phytoplankton taxa recorded over the three time periods and study sites. Phytoplankton were arranged into four functional groups: Cyanobacteria, diatoms (Bacillariophyceae), green algae (Chlorophyta), and flagellated Chromista (Dinoflagellates, Cryptomonads, and Chrysomonads). Phytoplankton counts were converted to relative abundance (#/m^3) and analyzed for 1970, 2003–2005, and 2017–2021. Taxa that accounted for less than 1% of the total abundance across the majority of the samples were not included in the analysis. Furthermore, the genera *Chroomonas*, *Cryptomonas*, and *Rhodomonas* (phylum Cryptista) were combined under the class Cryptophyceae, given their relatively small size and potential for misidentification. All other phytoplankton were identified to genus level. Mean phytoplankton relative abundances were calculated for each month during each of the sampling periods.

The resulting data set contained the top 15 most abundant taxa, which were analyzed using non-metric multidimensional scaling (NMDS) ordination, based on Bray-Curtis dissimilarity distances [35]. The ordinations were performed in R studio using the metaMDS function from the '*vegan*' package. NMDS ordinations are often applied to community abundance data to show distribution of dissimilarities between a large number of ecological samples [36]. The sampling time periods (1970, 2003–2005, and 2017–2021) and seasonality (May–September) were used to identify patterns in phytoplankton community composition. Simpsons diversity index was calculated to compare similarities among the three time periods. In addition, water quality parameter (surface temperature, total nitrogen, total phosphorus, and chloride) data from 1992 to 2021 were obtained from the Lake Champlain Long-term Monitoring database. A full list of phytoplankton taxa from each study period can be found in Appendix A.

3. Results

3.1 Long-term changes in phytoplankton composition and abundance

Phytoplankton community composition experienced shifts in dominance during the three time periods (**Figure 3**). In 1970, diatoms (40.92%) and protozoa (34.64%) were dominant within phytoplankton assemblages (**Figure 3**). Diatoms

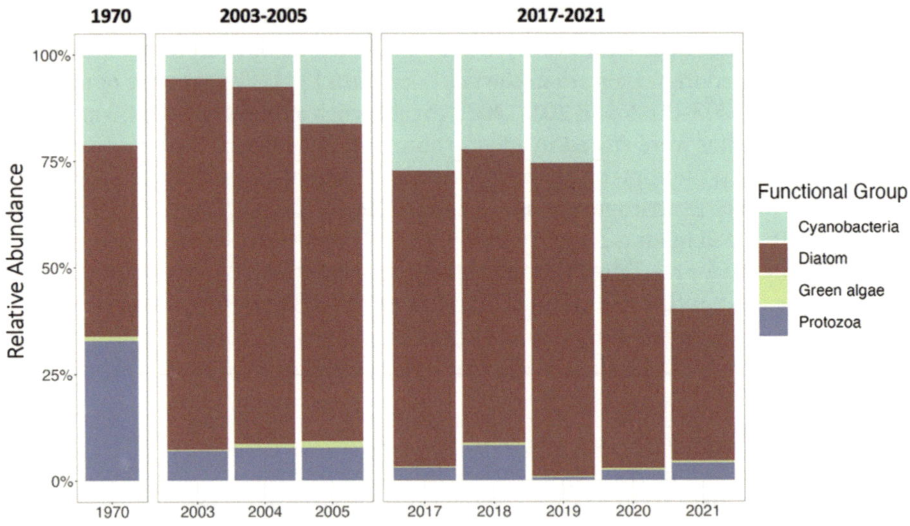

Figure 3.
Relative abundance of phytoplanktonic functional groups over the three time periods.

exceeded 50% of total abundance in 2003–2005 while protozoa Chromista were <10%. In 2017–2021, diatoms continued to be the main phytoplankton assemblage until 2020–2021 when Cyanobacteria exceeded 50% of total community abundance. A noticeable increase in cyanobacteria relative abundance occurred from 2017 to 2021. Phytoplankton assemblage composition was relatively similar within each study site during the 1970, 2003–2005, and 2017–2021 periods. Overall, between study periods, community shifts were observed, most notably a decline in protozoa abundance from 1970 to 2003–2005 and 2017–2021, while diatoms remained abundant throughout the entire study period and Cyanobacteria increased in the latter time periods.

Community patterns from the NMDS ordination showed differences in community composition between the three time periods (1970, 2003–2005, and 2017) (**Figure 4**). Particularly, there was a separation between phytoplankton assemblages from 1970 compared to 2003–2005 and 2017–2021. The phytoplankton assemblage shifted in composition primarily with respect to Cyanobacteria and diatom genera present in the community. This change is reflected in percent community similarity between the time periods (**Table 2**). Phytoplankton assemblages showed the lowest percent community similarity (~ 21.5%) between 1970 and 2003–2005; 2017–2021, indicating a major temporal shift in community composition. The highest similarity (40% or greater) was between samples from 2003 to 2005 and 2017–2021, suggesting there were fewer changes in overall community abundance between these two time periods. The similarity was highest within the 2017–2021 samples (68%) and in general remained high within all time periods. Likewise, there was a difference in Bray-Curtis similarity between years (1970, 2003–2005, and 2017–2021) and months during the study (**Table 3**).

Dolichospermum (formerly *Anabaena*), *Aphanizomenon* (Cyanobacteria), *Synedra* (Bacillariophyceae), *Schroederia* (Chlorophyta), and Cryptophyceae were more abundant in the 1970 samples in the ordination (**Figure 4**). The 2003–2005 and 2017–2021 sites were distributed away from 1970 sites and clustered moderately close together, illustrating the overlap between phytoplankton community structure between 2003

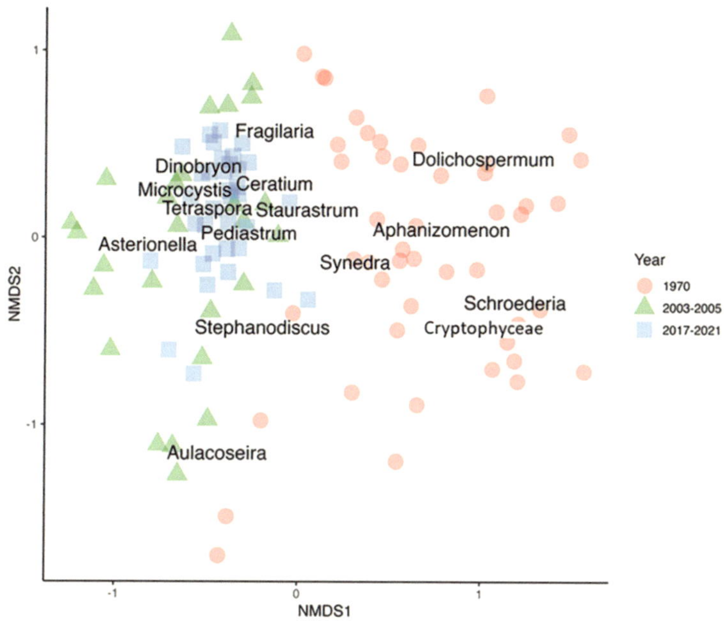

Figure 4.
NMDS ordination (using Bray-Curtis index of similarity) of the 15 most abundant phytoplankton during the three time periods (1970, 2003–2005, and 2017–2021), with a plot stress value of 0.189.

	1970	2003–2005	2017–2021
197 0	0.41413296	0.19921026	0.23245990
2003–2005		0.44345036	0.42240605
2017–2021			0.68061292

Table 2.
Average similarity (Bray-Curtis) for all sites and time periods.

Groups	PERMANOVA				
	Df	SS	R^2	F	p-Value
Years	2	9.478	0.36145	43.1764	0.001***
Site	6	0.8773	0.03346	1.3322	0.162
Month	4	5.1550	0.19659	11.7417	0.001***
Years: Site	12	1.2122	0.04623	0.9203	0.610
Years: Month	7	1.8123	0.06911	2.3588	0.002**
Site: Month	24	2.4165	0.09215	0.9173	0.658
Years: Site: Month	49	3.7345	0.14242	0.9724	0.781
Residual	14	1.5366	0.05860		

*Significance codes: 0 '***' 0.001 '**' 0.01 '*' 0.05.*

Table 3.
Permanova results using Bray-Curtis distance for the years, sites, and months.

and 2005 and 2017–2021 and the shifts in community structure from 1970. The phytoplankton that exhibited higher occurrences in samples collected during 2003–2005 and 2017–2021 include *Asterionella*, *Fragilaria* (Bacillariophyceae), *Microcystis* (Cyanobacteria), *Dinobryon* (Chrysophyceae), *Tetraspora* (Chlorophyceae), *Ceratium* (Dinophyceae), and *Pediastrum* (Chlorophyceae). Study site location did not show an effect on distribution of phytoplankton communities within or between time periods (**Table 3**), suggesting the observed community patterns and shifts across the time periods occurred lake wide.

Relative abundance plots were created using the ordination results to illustrate the shift in relative abundance for the dominant phytoplankton taxa with consideration of sites and time periods (**Figure 5**). The relative abundance for *Aphanizomenon* and *Dolichospermum* was the largest in 1970 samples lake wide (**Figure 5a,b**). These cyanobacteria remained present in the recent samples (2003–2005 and 2017–2021) but in lower abundance. The cyanobacterium *Microcystis* had higher abundance in

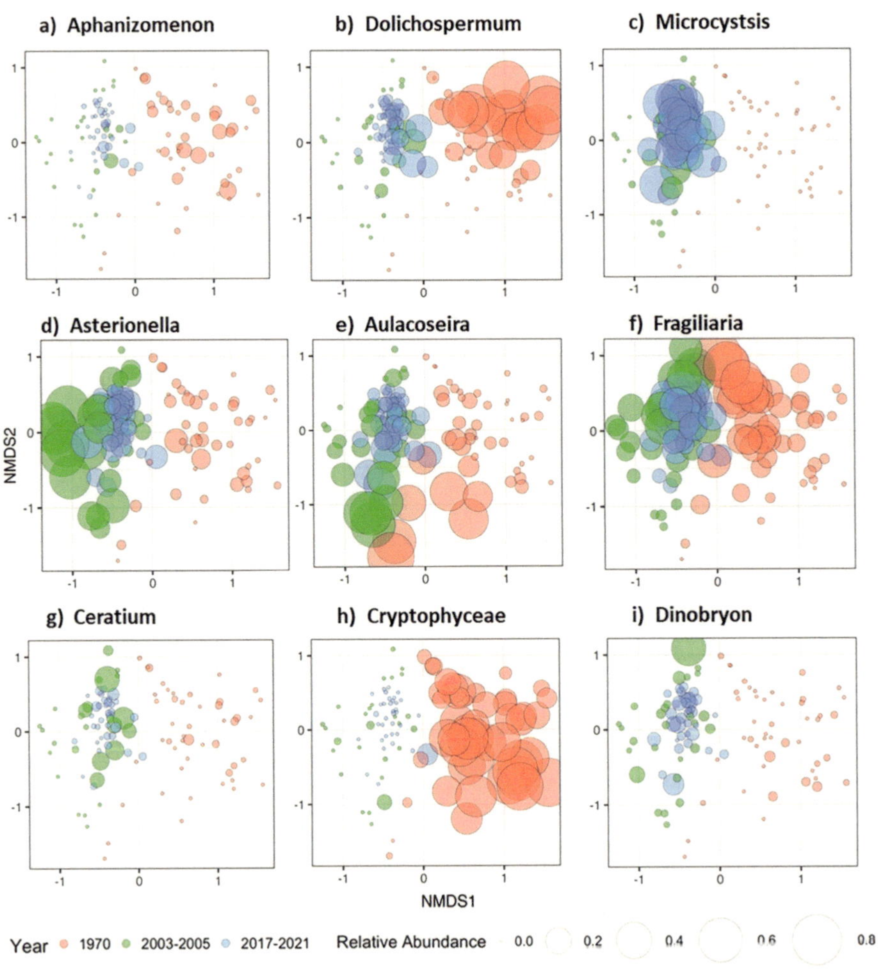

Figure 5.
Distribution of relative abundance between 1970 (red symbols), 2003–2005 (green symbols), and 2017–2021(blue symbols) for the dominant cyanobacteria (panels a, b, c), diatoms (panels d, e, f), and chromista (panels g, h, i) using NMDS ordination.

Lake Champlain during 2003–2005 and 2017–2021 (**Figure 5c**) and were absent in 1970 samples. Overall, Lake Champlain's Cyanobacteria community has exhibited a major temporal shift in taxa dominance from *Dolichospermum*, and *Aphanizomenon* (1970) to *Microcystis* (since the early 2000s) with the highest abundances observed in 2017–2021).

Asterionella, Aulacoseira, and *Fragilaria* were the dominant diatom genera throughout the three time periods (**Figure 5d–f**). *Asterionella* was lower in abundance in 1970, but increased in recent time periods, with particularly high levels in 2005. *Fragilaria* was overall the most abundant diatom in 1970 and 2017–2021 sampling period. Between 2017 and 2021, *Asterionella* (Bacillariophyceae) and *Aulacoseira* (Coscinodiscophyceae) increased in relative abundance during years that *Fragilaria* were less abundant in the samples. During 1970, Cryptophyceae were high in abundance for almost every sample (**Figure 5h**). In 2003–2005 and 2017–2021, *Ceratium* (Dinophyceae) and *Dinobryon* (Chrysophyceae) increased in abundance (**Figure 5g, i**). Overall, many taxa illustrated shifts in composition from 1970 to 2017–2021.

3.2 Seasonal succession of phytoplankton assemblages

We observed differences in the seasonal succession of phytoplankton assemblages, among the three time periods (**Figure 6**). In 1970, diatoms (Bacillariophyceae) and Chromista were the dominant early season phytoplankton groups until August. Within the diatom assemblage, *Aulacoseira* typically had high abundance in May, while *Fragilaria* generally increased in relative abundance in June and July for all time periods. Additionally, *Stephanodiscus* and *Synedra* accounted for a greater proportion of relative abundance within diatom assemblage in 1970 compared to recent sampling periods and were more abundant in the early season. Cryptophyceae were abundant throughout the sampling season of 1970 but showed a decline in September, following an increase in Cyanobacteria. *Dolichospermum* and *Aphanizomenon* were the dominant

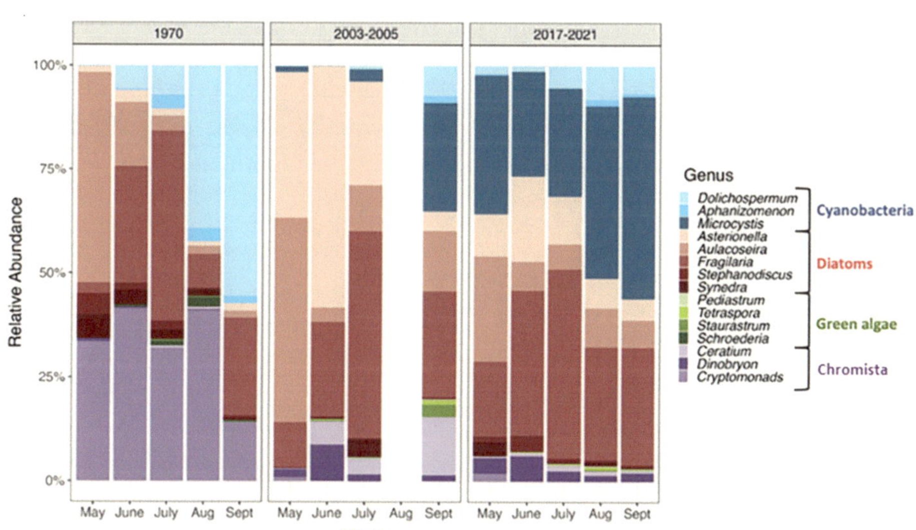

Figure 6.
Seasonal relative abundance for the top 15 most abundant phytoplankton between May and September for the three time periods.

Cyanobacteria in 1970 with notable increases in abundance during August. Green algae had greater relative abundances during July and August.

In 2003–2005, diatoms constituted a majority of the phytoplankton community composition and abundance. *Asterionella* increased from May to July during this period but were notably less abundant in the 1970 samples (**Figure 6**). *Aulacoseira* (Coscinodiscophyceae) and *Fragilaria* (Bacillariophyceae) maintained high relative abundance throughout the season. During September 2003–2005, there were increases in green algae, Cyanobacteria, and Chromista. A prominent switch in Cyanobacteria genera occurred from *Dolichospermum* dominance in 1970 to *Microcystis* in 2003–2005. Comparably, protozoa composition changed from primarily Cryptophyceae in 1970 to *Ceratium* and *Dinobryon*. Despite being a minor component of the overall relative abundance, green algae (Chlorophyceae) experienced a shift from *Schroederia* dominance in 1970 to *Tetraspora* and *Staurastrum*.

Between 2017 and 2021, we observed a rise in the abundance of Cyanobacteria, primarily *Microcystis*, which also occurred earlier in the growing season (**Figure 6**). *Microcystis* remained a dominant community member throughout 2017–2021 in Lake Champlain and was much less common in 1970 and 2003–2005. *Dolichospermum* and *Aphanizomenon* continued to exhibit lower abundance, with slight increases during August to September, similar to previous time periods. *Fragilaria* was the most abundant diatom for the growing season. Protozoa relative abundance generally decreased from 1970 to 2017–2021, with *Dinobryon* as the prevailing Chromista genera from 2017 to 2021.

The average surface water temperature (°C) at the seven study sites displayed an upward trend from 1992 to 2021 during the growing season (May–September) (**Figure 7, Table 4**). In May, there was a noticeable rise in surface temperature

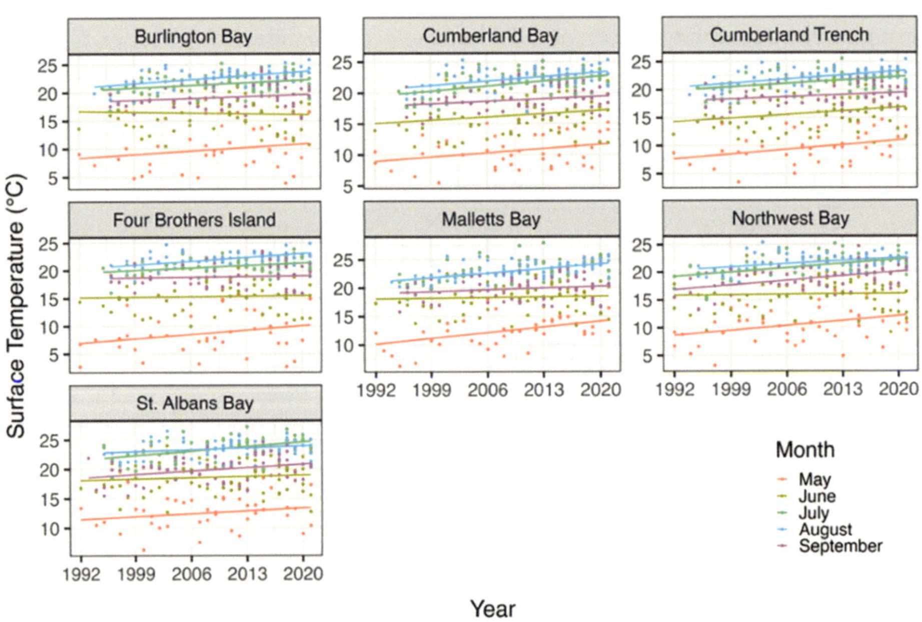

Figure 7.
Average surface temperature (°C) at 1 m for the seven sites from May to September from 1992 to 2021 with plotted linear regression lines.

Site name	1965–1969 Mean Secchi (m)	2003–2005 Mean Secchi (m)	2017–2021 Mean Secchi (m)	1970 Max temp (°C)	2003–2005 Max temp (°C)	2017–2021 Max temp (°C)
Northwest Bay	2.8	3.8	3.9	21.9	25.4	24.7
Four Brothers Island	4.7	5.5	4.9	22.1	24.7	24.9
Burlington Bay	4.4	5.4	4.5	23	24.9	25.8
Malletts Bay	4.0	5.3	4.3	25.3	26.2	25.8
Cumberland Bay	3.0	4.5	4.3	23.3	25.3	25.4
Cumberland Trench	5.4	5.2	5.1	23	24.8	25.4
St. Albans Bay	1.8	2.7	2.2	24.6	26.5	26.8

Maximum temperatures and mean Secchi depth for 1970 were based on Myers and Gruendling [1], while the 2003–2005 and 2017–2021 were obtained from the long-term monitoring (LTM) data.

Table 4.
Mean Secchi depth and maximum surface temperature for the seven study sites.

across the sites, with an average increase of 0.103°C. Malletts Bay, Northwest Bay, and Cumberland Trench had the highest recorded increases in May for the seven sites, during this time. In June, there were marginal rises in temperatures at six of the study sites. June and September exhibited overall average increases of 0.034 and 0.061°C, respectively. Comparatively, July and August had the same average increase of 0.094°C for all sites combined. Maximum water temperatures increased from 1970 to 2021 at all sites with increases up to 2.8°C over the 50-year period, primarily in the main lake (**Table 4**).

4. Discussion

4.1 Long-term changes in phytoplankton composition

Long-term shifts in phytoplankton community composition and abundance have occurred in Lake Champlain over a 50+ year time period from 1970 to 2021. In 1970, diatoms and protozoa were the dominant phytoplankton groups. Likewise, these two phytoplankton groups were predominant in samples from Lake Superior and Lake Erie, during this time [7]. Diatoms remained the predominant phytoplankton group in 2003–2005, with a decline in protozoa. This decline in protozoa likely occurred in the mid to late 1990s. Shambaugh et al. [5] found diatoms and cryptophytes flagellates were both dominant phytoplankton in 1991–1992. In 1993, zebra mussels invaded Lake Champlain and quickly achieved high abundance [23]. Since zebra mussels are efficient suspension feeders, this change in the community could have caused the shift in phytoplankton assemblages we observed between 1970 and the 2003–2005 period, similar to the impacts of zebra mussels on rotifer densities in Lake Champlain [18, 26]. A study in Lake Erie found that Dreissenidae (zebra mussel) invasion caused an increase in cyanobacterial abundance, while other phytoplankton assemblages were reduced [37]. During 1970 in Lake Champlain, water transparency was reported to decrease in spring and summer as phytoplankton proliferated [1]. Comparatively, our results found the mean Secchi depth (m) increased for all sites except Cumberland in subsequent time periods, which supports the suggestion that the zebra mussel "filtration effect" can induce changes in phytoplankton abundance and composition. During 2017–2021, Cyanobacteria and diatoms were the dominant groups in the phytoplankton community, but protozoa remained relatively less abundant.

One of the most notable results from our study was the long-term shift in dominant Cyanobacteria assemblage composition from *Dolichospermum* and *Aphanizomenon* in 1970 to *Microcystis* in 2003–2005 and 2017–2021. Similarly, in 1970, in Lake Erie, *Dolichospermum* and *Aphanizomenon flos-aquae* (formerly *Anabaena flos-aquae*) were common Cyanobacteria that formed summer blooms [9]. Both of these Cyanobacteria are nitrogen fixers, which provides a competitive advantage under nitrogen-limited conditions relative to other phytoplankton [38]. By the mid-1990s Cyanobacteria blooms occurred more frequently in Lake Erie, but *Microcystis aeruginosa* had become the dominant species [39]. The general shift in dominant Cyanobacteria genera to *Microcystis* in numerous freshwater lakes and rivers worldwide suggests similar underlying factors influencing the change [39, 40]. The highest abundance of *Microcystis* observed in our study was during 2020–2021, following invasion of "fishhook waterflea" (*Cercopagis pengoi*) in Lake Champlain in 2018–2019. This suggests possible planktonic food web interactions following introduction of a top planktonic predator may have resulted in increased niche availability for *Microcystis*. These possible trophic interactions require more study.

The increasing predominance of *Microcystis* in Lake Champlain and the Great Lakes could be due to a combination of factors including increased invasive species mediated interactions, water temperature increases and altered nutrient dynamics [15]. Warm water temperatures are likely to be the factor influencing *Microcystis* bloom formations in Lake Champlain, with large blooms forming above 25°C, a threshold more commonly reached in recent decades [41–43]. The shift in dominance could be a result of *Microcystis* having a higher optimal temperature range compared to *Dolichospermum* and *Aphanizomenon*. Experiments performed in Lake Champlain demonstrated that elevated water temperatures yielded significantly higher growth rates of toxic and non-toxic *Microcystis* [43]. Additionally, nitrogen and phosphorus enrichment lead to notably increased growth rates of toxic strains of *Microcystis* [43]. One study noted that *Microcystis* has an optimal N:P ratio of 32:1 [40], while another study which noted high abundance of microcystin was found in <23 N:P ratio environments [44]. Paerl and Huisman [45] found that *Microcystis* blooms became more prevalent with a combination of increased temperatures, nutrient inputs, and CO_2 levels. *Microcystis* blooms have been noted to reduce nitrogen in the water column [46].

The most common diatoms for our study sites during 1970 were *Aulacoseira*, *Fragilaria, Stephanodiscus,* and *Synedra*. Many of the species in these genera are considered eutrophic or mesotrophic indicators [47]. We observed a general shift toward a high abundance of *Asterionella* in 2003–2005, which has been noted as an oligotrophic or mesotrophic diatom [47]. Comparatively, in Lake Erie, *Asterionella* was dominant until 1950 when nutrient-dominant diatoms like *Aulacoseira* and *Fragilaria*

DOI: http://dx.doi.org/10.5772/intechopen.1007276

increased [47]. In Lake Champlain during 2017–2021, *Asterionella* decreased in abundance compared to other time periods, while *Fragilaria* became the dominant diatom. *Fragilaria* have been noted to increase with higher inputs of phosphorus and a superior competitor for silica compared to *Asterionella* [47]. In addition, *Fragilaria* has an optimal temperature of 10–30°C, which is higher than most diatoms found in Lake Champlain (see Appendix B) [48]. In another study from Lake Michigan, a decrease in *Asterionella* was observed while *Stephanodiscus* increased with phosphorus inputs [49]. The transition in phytoplankton assemblages to taxa better adapted to warmer water is a common occurrence throughout the Great Lakes and Lake Champlain and other lakes worldwide.

Cryptomonads were the dominant perennial protozoa in 1970. Cryptomonads and Cyanobacteria have a red accessory pigment phycoerythrin, which allows for high photosynthetic absorption rate and ability to grow at greater depths [50]. A mesocosm experiment found that warming water caused a relative decrease in Cryptophyceae abundance [51]. These relatively small phytoplankton are often found in deep water column of oligotrophic lakes [52]. A general decline was observed for Cryptophyceae from 1970 to 2003–2005, which could be a result of sample collection depth and increased water temperature. However, Naddafi et al. [53] found that zebra mussels reduced biomass of Cryptophyceae. Likewise, zooplankton grazing pressure have been reported to select Cryptophyceae as prey, which could have impacted their abundance [47]. In 2003–2005, the dominant Chromista was *Ceratium* (Dinophyceae), which commonly inhabits mesotrophic to eutrophic conditions during stable stratification [47]. Comparably, during 2017–2021, protozoa were overall less abundant with *Dinobryon* (Chrysophyceae) as the most common.

4.2 Seasonal succession of phytoplankton

Phytoplankton community composition and abundance changed throughout the growing season during 1970, 2003–2005, and 2017–2021. In freshwater ecosystems, water quality variables such as temperature, nutrients, and sunlight strongly influence phytoplankton dynamics [47]. Seasonal succession of phytoplankton assemblages in dimictic lakes generally follows a pattern of high abundance of diatoms in spring turnover due to the readily available silica, essential for the cell walls of diatoms [54], and relatively low temperatures and light intensity [55]. In summer, green algae (Chlorophyta), Cryptophyceae, and Cyanobacteria typically become dominant within phytoplankton assemblage [55]. Cyanobacteria are typically the most abundant in late summer because they have a higher optimal temperature compared to other phytoplankton assemblages, along with some Cyanobacteria being grazing resistant [39]. Additionally, Cyanobacteria are better equipped for handling nutrient depletion in the epilimnion due to lower nutrient demands, and certain genera have the capability to perform nitrogen fixation [56]. Diatoms typically return as the dominant phytoplankton assemblage in fall turnover [55].

Seasonal succession of phytoplankton followed a similar pattern in 1970 and 2003–2005 sampling periods in Lake Champlain. Diatoms were the main community assemblage from May to July. Green algae were most abundant in June and July in 1970 samples but now appear at higher density later in the field season in recent years. Filter-feeding zooplankton has been noted to preferentially graze on green algae, causing their abundance to decline [57]. Davis et al. [43] found Cyanobacteria blooms occurred in late summer during 2006 in Missisquoi Bay (Lake Champlain), with *Dolichospermum* as the dominant Cyanobacteria in late summer, then transitioning to *Microcystis* dominance.

During 2017–2021, this seasonal pattern changed, with Cyanobacteria abundance increasing earlier in the season (May–June), while other phytoplankton assemblages decreased in relative abundance. Cyanobacteria blooms have been shown to outcompete other phytoplankton groups in warm water, which could have caused a decline in some taxa during 2017–2021 [54]. Surface water temperature (°C) has exhibited a continuous rise at all seven sites since 1992 and is most likely the main driver for the increased Cyanobacteria abundance earlier in the growing season in recent years.

5. Conclusion

Long-term monitoring of phytoplankton community composition and seasonal abundance is useful to provide a better understanding of long time scale patterns in freshwater lakes. In this study, changes in phytoplankton community composition and abundance occurred over a 50-year time scale. Diatoms (Bacillariophyceae) and Chromista were the predominant phytoplankton assemblages in 1970 but shifted to cyanobacteria and diatoms in 2003–2005 and 2017–2021. Cyanobacteria displayed higher abundance earlier in the growing season during 2017–2021 compared to previous sampling periods. Notably, *Dolichospermum* and *Aphanizomenon* were dominant in 1970 but shifted to *Microcystis* dominance in 2003–2005 and 2017–2021, illustrating that phytoplankton that are adapted to warmer waters may replace previously dominant phytoplankton genera over time. Shifts in community composition also occurred among diatoms and protozoa from 1970 to 2021. Overall, our results suggest that climate change and invasive species-mediated impacts (e.g., zebra mussel) are likely the main causes for shifts in phytoplankton assemblage in Lake Champlain.

Acknowledgements

We would like to thank Lake Champlain Basin Program (funded by USEPA) for supporting Lake Champlain Long-Term Monitoring Project (LTMP) collected in 2017-2021. Additional support was provided by Lake Champlain Research Institute at SUNY Plattsburgh for data collected in 2003-2005.

Appendices A

List of phytoplankton genera identified for each time period

See **Table A1**.

1970	2003–2005	2017–2021
Cyanobacteria	Cyanobacteria	Cyanobacteria
Aphanizomenon	*Aphanizomenon*	*Aphanizomenon*
Coelosphaerium	*Aphanocapsa*	*Aphanocapsa*
Dolichospermum	*Coelosphaerium*	*Coelosphaerium*
Pseudomonas	*Cylindrospermum*	*Cylindrospermum*
	Dolichospermum	*Dolichospermum*
	Eucapsis	*Gloeotrichia*
	Gloeocapsa	*Merismopedia*
	Gloeotrichia	*Microcystis*
	Microcystis	*Snowella*
	Nostoc	
	Oscillatoria	
	Snowella	
	Spirulina	
Green Algae	Green Algae	Green algae
Actinastrum	*Ankistrodesmus*	*Actinastrum*
Ankistrodesmus	*Botryococcus*	*Ankistrodesmus*
Chlamydomonas	*Coelastrum*	*Coelastrum*
Coelastrum	*Dictyosphaerium*	*Dictyosphaerium*
Dictyosphaerium	*Eudorina*	*Eudorina*
Eudorina	*Hematococcus*	*Kirchneriella*
Oocystis	*Kirchneriella*	*Pediastrum*
Pediastrum	*Mougeotia*	*Sphaerocystis*
Scenedesmus	*Oedogonium*	*Spirogyra*
Schroederia	*Oocystis*	*Tetraspora*
Tetrastrum	*Pediastrum*	*Crucigenia*
	Scenedesmus	*Microactinium*
	Sphaerocystis	*Volvox*
	Spirogyra	
	Tetraspora	
	Volvox	
Desmid	Desmid	Desmid
Closterium	*Closterium*	*Closterium*
Cosmarium	*Cosmarium*	*Staurastrum*
Staurastrum	*Staurastrum*	*Staurodesmus*
	Staurodesmus	*Xanthidium*
	Xanthidium	*Xanthidium*
Diatom	Diatom	Diatom
Asterionella	*Amphora*	*Asterionella*

1970	2003–2005	2017–2021
Fragilaria	*Asterionella*	*Cyclotella*
Aulocoseira	*Cyclotella*	*Diatoma*
Navicula	*Diatoma*	*Fragilaria*
Nitzschia	*Fragilaria*	*Aulocoseira*
Stephanodiscus	*Frustulia*	*Navicula*
Synedra	*Aulocoseira*	*Stephanodiscus*
Tabellaria	*Navicula*	*Synedra*
Suriella	*Stephanodiscus*	*Tabellaria*
	Synedra	*Suriella*
	Tabellaria	
Chromista	Chromista	Chromista
Ceratium	*Acanthocystis*	*Acanthocystis*
Chroomonas/Cryptomonas	*Actinosphaerium*	*Actinosphaerium*
Chrysococcus	*Ceratium*	*Ceratium*
Dinobryon	*Chroomonas/Cryptomonas*	*Chroomonas/cryptomonas*
Glenodinium	*Difflugia*	*Difflugia*
Gymnodinium	*Dinobryon*	*Dinobryon*
Mallomonas	*Euglena*	*Euglena*
Peridinium	*Euglypha*	*Euglypha*
Rhodomonas	*Mallomonas*	*Mallomonas*
Synura	*Synura*	*Synura*
Rhizosolenia	*Trachelomonas*	*Trachelomonas*
		Vorticella
Miscellaneous		
Coccoid Green		

Table A1.

B. Ecological traits of the top 15 phytoplankton taxa in our study

See **Tables B1, B2**.

DOI: http://dx.doi.org/10.5772/intechopen.1007276

Taxon	Order	Optimal temp. range	Motile (Y/N)	Modes of nutrition	Light	Nutrient status	Turbulent mixing preference	Season dominate	References
Cyanobacteria (Cyaniohyceae)									
Aphanizomenon	Nostocales	23–29°C (*A. flosaquae)*	Y	Autotrophic	Moderate	Mesotrophic Eutrophic	Low	Summer	[32, 58–60]
Dolichospermum (Anabaena)	Nostocales	23–30°C	Y	Autotrophic	High	Eutrophic	Low	Summer	[61–63]
Microcystis	Chroococcales	24–34°C (M. *aeruginosa)*	Y	Autotrophic	Moderate	Eutrophic Hypereutrophic	Low	Summer	[54, 61, 64, 65]
Diatoms (Bacillariophyceae)									
Asterionella	Fragilariales	6–15°C (A. *formosa)*	N	Autotrophic	Low	Mesotrophic	Moderate	Spring	[47, 61, 66, 67]
Aulacoseira	Aulacoseirales	6–20°C (*A. ambigua)*	N	Autotrophic	Low	Mesotrophic Eutrophic	Moderate	Late-summer, Spring	[32, 61, 67]
Fragilaria	Fragilariales	10–30°C (*F. nanana)*	N	Autotrophic	Varies	Eutrophic	Low	Summer	[61, 66–69]
Stephanodiscus	Thalassiosirales	5–10°C	N	Autotrophic	Low	Eutrophic	Moderate	Winter, Early Spring	[7, 47, 70]
Synedra	Fragilariales	13–15°C	N	Autotrophic	Low	Mesotrophic Eutrophic	Moderate	Winter	[47, 71]

Table B1.
Generalized characteristic traits of the top 15 phytoplankton taxa in our study.

Taxon	Order	Optimal temp. range	Motile (Y/N)	Modes of nutrition	Light	Nutrient status	Turbulent mixing preference	Season dominate	References
Green Algae									
Pediastrum	Sphaeropleales	15–26°C	N	Autotrophic	Moderate	Mesoeutrophic	Low	Summer	[47, 72, 73]
Tetraspora	Chlamydomonadales	Cool to Moderate	N	Autotrophic	High	Mesotrophic	Relatively Low	Mid-summer	[47, 74, 75]
Schroederia	Sphaeropleales	N/A	N	Autotrophic	N/A	Oligotrophic	Low	Summer	[76–78]
Desmid									
Staurastrum	Desmidiales	14–25°C *(S. cingulum)*	N	Autotrophic	Low/ Mild	Mesotrophic Oligotrophic	Moderate	Summer	[59, 61, 79]
Chromista									
Ceratium	Gonyaulacales	18–28°C *(C furca)*	Y	Mixotrophic	Low	Mesotrophic Eutrophic	Low	Summer	[47, 57, 80, 81]
Dinobryon	Chromulinales	9–18°C	Y	Mixotrophic	Moderate	Oligotrophic	Low	Winter, Spring	[47, 61, 82]
Cryptophyceae	Cryptomonadales	16–20°C	Y	Mixotrophic	Low	Mesotrophic	Moderate	Winter, Spring	[47, 83, 84]

Table B2.
Generalized characteristic traits of the top 15 phytoplankton taxa in our study (Continued).

Author details

Timothy B. Mihuc*, Kayleen Snyder, Zachary Cutter and Shelly Bouyea
Lake Champlain Research Institute, SUNY Plattsburgh, Plattsburgh, NY, USA

*Address all correspondence to: timothy.mihuc@plattsburgh.edu

References

[1] Myers G, Gruendling G. Limnology of Lake Champlain. Burlington, VT: Lake Champlain Basin Study; 1979

[2] Stickney M, Hickey C, Hoerr R. Lake Champlain basin program: Working together today for tomorrow. Lakes and Reservoirs: Science, Policy and Management for Sustainable Use. 2001;**6**:217-223

[3] Meals DW, Budd LF. Lake Champlain basin nonpoint source phosphorus assessment. Journal of the American Water Resources Association. 1998;**34**:251-265

[4] Muenscher WG. Plankton studies in the Lake Champlain watershed. In: A Biological Survey of the Lake Champlain Watershed. NY State Conservation Department, Supplement to the 19th Annual Report. Albany, NY: State of New York Conservation Department; 1929

[5] Shambaugh AD, Duchovnay A, McIntosh A. A survey of Lake Champlain's plankton. In: Manley TO, Manley PL, editors. Lake Champlain in Transition: From Research toward Restoration, Water Science and Application. Vol. 1. Washington, D.C.: American Geophysical Union; 1999. pp. 323-340

[6] Reavie ED, Barbiero RP, Allinger LE, Warren GJ. Phytoplankton trends in the Great Lakes, 2001-2011. Journal of Great Lakes Research. 2014;**40**:618-639

[7] Munawar M, Munawar IF. A lakewide study of phytoplankton biomass and its species composition in Lake Erie, April–December 1970. Journal of the Fisheries Research Board of Canada. 1976;**33**:581-600

[8] Munawar M, Munawar IF. Phytoplankton of lake superior 1973. Journal of Great Lakes Research. 1978;**4**:415-442

[9] Makarewicz JC. Phytoplankton and Zooplankton in Lakes Erie, Huron, and Michigan: 1985. Chicago, IL: US Environmental Protection Agency, Great Lakes National Program Office; 1991

[10] Millie DF, Fahnenstiel GL, Dyble Bressie J, Pigg RJ, Rediske RR, Klarer DM, et al. Late-summer phytoplankton in western Lake Erie (Laurentian Great Lakes): Bloom distributions, toxicity, and environmental influences. Aquatic Ecology. 2009;**43**:915-934

[11] McAllister DE, Hamilton AL, Harvey B. Global freshwater biodiversity: Striving for the integrity of freshwater ecosystems. Sea Wind: Bulletin of Ocean Voice International. 1997;**11**(3):6-103

[12] Woodward G, Perkins DM, Brown LE. Climate change and freshwater ecosystems: Impacts across multiple levels of organization. Philosophical Transactions of the Royal Society B: Biological Sciences. 2010;**365**:2093-2106

[13] O'Reilly CM, Sharma S, Gray DK, Hampton SE, Read JS, Rowley RJ, et al. Rapid and highly variable warming of lake surface waters around the globe. Geophysical Research Letters. 2015;**42**:10-773

[14] Stager C, Thill M. Climate change in the Champlain Basin: What natural resource managers can expect and do. The Nature Conservancy. 2010:44

[15] Smeltzer E, Shambaugh AD, Stangel P. Environmental change in Lake Champlain revealed by long-term monitoring. Journal of Great Lakes Research. 2012;**38**:6-18

[16] Sivarajah B, Rühland KM, Labaj AL, Paterson AM, Smol JP. Why is the relative abundance of *Asterionella formosa* increasing in a boreal shield lake as nutrient levels decline? Journal of Paleolimnology. 2016;**55**:357-367

[17] Kruk C, Peeters ETHM, Van Nes EH, Huszar VLM, Costa LS, Scheffer M. Phytoplankton community composition can be predicted best in terms of morphological groups. Limnology and Oceanography. 2011;**56**:110-118

[18] Mihuc TB, Dunlap F, Binggeli C, Myers L, Pershyn C, Groves A, et al. Long-term patterns in Lake Champlain's zooplankton: 1992-2010. Journal of Great Lakes Research. 2012;**38**:49-57

[19] Gulka JE, Borah A, Koch C, Bering J, Shawler A, Sweeney M. Preventing the spread of aquatic invasive species into Lake Champlain. Lake Champlain Basin Program. 2013:1-92

[20] Ricciardi A, Rasmussen JB. Extinction rates of North American freshwater fauna. Conservation Biology. 1999;**13**:1220-1222

[21] Facey DE, Marsden JE, Mihuc TB, Howe EA. Lake Champlain 2010: A summary of recent research and monitoring initiatives. Journal of Great Lakes Research. 2012;**38**:1-5

[22] Balk N, Brett P, Good S, Jenness L, Skawinski P, Vennie-Vollrath E. The Lake Champlain basin aquatic invasive species guide version 3 2022. Lake Champlain Basin Program. 2023;**3**:1-37

[23] Stangel P, Shambaugh A. Lake Champlain 2003 Zebra Mussel Monitoring Program. Waterbury, Vermont: Vermont Department of Environmental Conservation; 2004

[24] Miller EB, Watzin MC. The effects of zebra mussels on the lower planktonic foodweb in Lake Champlain. Journal of Great Lakes Research. 2007;**33**:407-420

[25] Idrisi N, Mills EL, Rudstam LG, Stewart DJ. Impact of zebra mussels (*Dreissena polymorpha*) on the pelagic lower trophic levels of Oneida Lake, New York. Canadian Journal of Fisheries and Aquatic Sciences. 2001;**58**:1430-1441

[26] Mihuc TB, Recknagel F. Lessons from bioinvasion of Lake Champlain, USA. In: Ecological Informatics: Data Management and Knowledge Discovery. Cham, CH: Springer; 2018. pp. 401-414

[27] Cutter ZA, Mihuc TB, Myers LW. Invasion of *Bythotrephes longimanus* and *Cercopagis pengoi* in Lake Champlain: Impacts on the native zooplankton community. Diversity. 2023;**15**:1112. DOI: 10.3390/d15111112

[28] Pagnucco KS, Maynard GA, Fera SA, Yan ND, Nalepa TF, Ricciardi A. The future of species invasions in the Great Lakes-St. Lawrence River basin. Journal of Great Lakes Research. 2015;**41**:96-107

[29] Rohwer RR, Hale RJ, Vander Zanden MJ, Miller TR, McMahon KD. Species invasions shift microbial phenology in a two-decade freshwater time series. Proceedings of the National Academy of Sciences. 2023;**120**(11):1-8

[30] O'Malley BP, Rudstam LG, Watkins JM, Holda TJ, Weidel BC. Effects of food web changes on *Mysis diluviana* diet in Lake Ontario. Journal of Great Lakes Research. 2017;**43**(5):813-822

[31] Lovett GM, Burns DA, Driscoll CT, Jenkins JC, Mitchell MJ, Rustad L, et al. Who needs environmental monitoring? Frontiers in Ecology and the Environment. 2007;**5**:253-260

[32] Henson EB, Gruendling G. The Trophic Status and Phosphorus Loadings of Lake Champlain. Corvalis, OR: Environmental Protection Agency; 1977

[33] Vermont Department of Environmental Conservation (VT DEC) and New York Department of Environmental Conservation (NY DEC). Long-Term Water Quality and Biological Monitoring. 2021. Available from: https://anrweb.vermont.gov/dec/dec/LongTermMonitoringLakes.aspx

[34] Picotte A. Lake Champlain Lay Monitoring Report. Waterbury, VT: Agency of Natural Resources Department of Environmental Conservation Water Quality Division; 2008

[35] Oksanen J, Blanchet G, Kindt R, Legendre P, Minchin PR, Oh'ara RB, et al. Package 'vegan'. Community Ecology Package, Version. 2013;**2**(9):1-295

[36] Minchin P. An evaluation of the relative robustness of techniques for ecological ordination. Theory and Models in Vegetation Science: Proceedings of Symposium, Uppsala. 1987;**69**:89-107

[37] Thomas CM. A Cost-Benefit Analysis of Preventative Management for Zebra and Quagga Mussels in the Colorado-Big Thompson System. Fort Collins, CO: Colorado State University; 2010

[38] Huisman J, Codd GA, Paerl HW, Ibelings BW, Verspagen JMH, Visser PM. Cyanobacterial blooms. Nature Reviews Microbiology. 2018;**16**:471-483

[39] Vanderploeg HA, Liebig JR, Carmichael WW, Agy MA, Johengen TH, Fahnenstiel GL, et al. Zebra mussel (*Dreissena polymorpha*) selective filtration promoted toxic *Microcystis* blooms in Saginaw Bay (Lake Huron) and Lake Erie. Canadian Journal of Fisheries and Aquatic Sciences. 2001;**58**:1208-1221

[40] Liu C, Shi X, Fan F, Wu F, Lei J. N:P ratio influences the competition of Microcystis with its picophytoplankton counterparts, *Mychonastes* and *Synechococcus*, under nutrient enrichment conditions. Journal of Freshwater Ecology. 2019;**34**:445-454

[41] Boyer GL, Satchwell MF, Shambaugh A, Watzin M, Mihuc TB, Rosen B. The occurrence of cyanobacterial toxins in Lake Champlain waters. In: Manley T, Manley P, Mihuc TB, editors. Lake Champlain: Partnerships and Research in the New Millennium. New York Press: Kluwer Academic/Plenum Publishers; 2004. pp. 241-257

[42] Mihuc TBC, Pershyn S, Thomas G, Boyer M, Satchwell J, Jones EA, et al. Cyanobacteria and the sixth great lake: Community dynamics of toxic algal blooms in Lake Champlain, USA. Verhandlungen des Internationalen Verein Limnologie. 2008;**30**:312-317

[43] Davis TW, Berry DL, Boyer GL, Gobler CJ. The effects of temperature and nutrients on the growth and dynamics of toxic and non-toxic strains of microcystis during cyanobacteria blooms. Harmful Algae. 2009;**8**:715-725

[44] Orihel DM, Bird DF, Brylinsky M, Chen H, Donald DB, Huang DY, et al. High microcystin concentrations occur only at low nitrogen-to-phosphorus ratios in nutrient-rich Canadian lakes. Canadian Journal of Fisheries and Aquatic Sciences. 2012;**69**:1457-1462

[45] Paerl HW, Huisman J. Climate change: A catalyst for global expansion of harmful cyanobacterial blooms. Environmental Microbiology Reports. 2009;**1**:27 37

[46] Shen Y, Huang Y, Hu J, Li P, Zhang C, Lei L, et al. The nitrogen reduction in eutrophic water column driver

by microcystis blooms. Journal of Hazardous Materials. 2020;**385**:121578

[47] Sheath R, Wehr J. Freshwater Algae of North America: Ecology and Classification. Washington, D.C.: Academia Press; 2003

[48] Zhang Y, Peng C, Wang Z, Zhang J, Li L, Huang S, et al. The species-specific responses of freshwater diatoms to elevated temperatures are affected by interspecific interactions. Microorganisms. 2018;**6**:82

[49] Makarewicz JC, Baybutt RI. Long-term (1927-1978) changes in the phytoplankton community of Lake Michigan at Chicago. Bulletin of the Torrey Botanical Club. 1981;**108**:240

[50] Goodwin TW. Carotenoids and biliproteins. Algal physiology and biochemistry. Botanical Monographs. 1974;**10**:176-205

[51] Moss B, Atkinson D, Collings SE, Eaton JW, Gill AB, Harvey I, et al. How important is climate? Effects of warming, nutrient addition and fish on phytoplankton in shallow lake microcosms. Journal of Applied Ecology. 2003;**40**:782-792

[52] Nauwerk A. Das Phytoplankton des Latnjajaure. 1954-55. Schweizerische Zeitschrift für Hydrologie. 1968;**30**:188-216

[53] Naddafi R, Pettersson K, Eklöv P. The effect of seasonal variation in selective feeding by zebra mussels (*Dreissena polymorpha*) on phytoplankton community composition. Freshwater Biology. 2007;**52**:823-842

[54] Wehr JD, Sheath RG, Kociolek JP. Freshwater Algae of North America: Ecology and Classification. 2nd ed. Amsterdam; Boston: Elsevier/AP, Academic Press is an imprint of Elsevier; 2015

[55] Levine S, Shambaugh A, Pomeroy S, Braner M. Phosphorus, nitrogen, and silica as controls on phytoplankton biomass and species composition in Lake Champlain (USA-Canada). Journal of Great Lakes Research. 1997;**23**:131-148

[56] Carr NG, Whitton BA. The biology of cyanobacteria. Vol. 19. Oxford: Blackwell Scientific; 1982. pp. 87-104

[57] Reynolds C. The Ecology of Phytoplankton. Cambridge, UK: Cambridge University Press; 1984

[58] Gentile JH, Maloney TE. Toxicity and environmental requirements of a strain of *Aphanizomenon flos-aquae* (L.) Ralfs. Canadian Journal of Microbiology. 1969;**15**:165-173

[59] Reynolds N. Seasonal variations in Staurastrum Paradoxum Meyen. New Phytologist. 1940;**39**:86-89

[60] Tsujimura S, Ishikawa K, Tsukada H. Effect of temperature on growth of the cyanobacterium Aphanizomenon flos-aquae in Lake Biwa and Lake Yogo. Phycological Research. 2001;**49**:275-280

[61] Rawson DS. Algal indicators of trophic lake types. Limnology and Oceanography. 1956;**1**:18-25

[62] Nobel WTP, Huisman J, Snoep JL, Mur LR. Competition for phosphorus between the nitrogen-fixing cyanobacteria *anabaena* and aphanizomenon. FEMS Microbiology Ecology. 2006;**24**:259-267

[63] Kłodawska K, Bujas A, Turos-Cabal M, Żbik P, Fu P, Malec P. Effect of growth temperature on biosynthesis and accumulation of carotenoids in cyanobacterium *anabaena*

sp. PCC 7120 under diazotrophic conditions. Microbiological Research. 2019;**226**:34-40

[64] Imai H, Chang K-H, Kusaba M, S. -i. Nakano. Temperature-dependent dominance of *Microcystis* (Cyanophyceae) species: *M. Aeruginosa* and *M. Wesenbergii*. Journal of Plankton Research. 2008;**31**:171-178

[65] Litchman E, de Tezanos Pinto P, Klausmeier CA, Thomas MK, Yoshiyama K. Linking traits to species diversity and community structure in phytoplankton. Hydrobiologia. 2010;**653**:15-28

[66] Smol J, Stoermer E. The Diatoms Applications for the Environmental and Earth Sciences. 2nd ed. Cambridge, UK: Cambridge University Press; 2010

[67] Zhang X, Xu X, Su L, Shen Z. Ecological responses of the diatom species *Asterionella formosa* to climate change and resource availability in a shallow eutrophic lake of Chinese loess plateau. Fundamental and Applied Limnology. 2023;**196**:93-105

[68] Hartig JH. Factors contributing to the development of Fragilaria Crotonesis Kitton "pulses" in the Pigeon bay waters of Western Lake Erie [Electronic theses and dissertations]. Windsor, ON: University of Windsor; 1985

[69] Bhutiani R, Khanna DR, Chandra KS. Light-limited population dynamics of phytoplankton: Modeling light and depth effects. The Environmentalist. 2009;**29**:93-105

[70] Jung SW, Kwon OY, Lee JH, Han M-S. Effects of water temperature and silicate on the winter blooming diatom *Stephanodiscus hantzschii* (Bacillariophyceae) growing in eutrophic conditions in the lower Han River, South Korea. Journal of Freshwater Ecology. 2009;**24**:219-226

[71] Sulastri CH, Santoso AB. Phytoplankton composition and the occurrence of cyanobacterial bloom in Lake Maninjau, Indonesia. IOP Conference Series: Earth and Environmental Science. 2019;**380**:012020

[72] Jankovská V, Komárek J. Indicative value of pediastrum and other coccal green algae in palaeoecology. Folia Geobotanica. 2000;**35**:59-82

[73] Lenarczyk J. The Algal Genus *Pediastrum* Meyen (Chlorophyta) in Poland. Krakow, Poland: W. Szafer Institute of Botany; 2014

[74] Jacobs V. The Algal Flora of the East Gallatin River as a Contribution to the Algae of Montana. Bozeman, MT: Montana State University; 1950

[75] Richter D, Matuła J, Pietryka M. The northernmost populations of *Tetraspora gelatinosa* (Chlorophyta) from spitsbergen. Polish Polar Research. 2014;**35**:521-538

[76] Li F, Zhang H, Zhu Y, Xiao Y, Chen L. Effect of flow velocity on phytoplankton biomass and composition in a freshwater lake. Science of the Total Environment. 2013;**447**:64-71

[77] Abonyi A, Descy JP, Borics G, Smeti E. From historical backgrounds towards the functional classification of river phytoplankton sensu Colin S. Reynolds: What future merits the approach may hold? Hydrobiologia. 2021;**848**:131-142

[78] Hu B, Zhou J, Dong J, Yang H, Yu G, Hong Y. Association of algae diversity and *Hyriopsis schlegelii* growth in mixed fish-mussel aquaculture. Algal Research. 2022;**65**:102736

[79] Butterwick C, Heaney SI, Talling JF. Diversity in the influence of temperature on the growth rates of freshwater algae, and its ecological relevance. Freshwater Biology. 2004;**50**:291-300

[80] Baek SH, Shimode S, Kikuchi T. Growth of dinoflagellates, *Ceratium furca* and *Ceratium fusus* in Sagami Bay, Japan: The role of temperature, light intensity and photoperiod. Harmful Algae. 2008;**7**:163-173

[81] Baek SH, Shin HH, Choi HW, Shimode S, Hwang OM, Shin K, et al. Ecological behavior of the dinoflagellate *Ceratium furca* in Jangmok harbor of Jinhae Bay, Korea. Journal of Plankton Research. 2011;**33**:1842-1846

[82] Heinze A, Truesdale C, DeVaul S, Swinden J, Sanders R. Role of temperature in growth, feeding, and vertical distribution of the mixotrophic chrysophyte Dinobryon. Aquatic Microbial Ecology. 2013;**71**:155-163

[83] Hammer A, Pitchford J. Mixotrophy, allelopathy and the population dynamics of phagotrophic algae (cryptophytes) in the Darss Zingst Bodden estuary, southern Baltic. Marine Ecology Progress Series. 2006;**328**:105-115

[84] Cunha Pereira H, Allott N, Coxon C, Naughton O, Johnston P, Gill L. Phytoplankton of turloughs (seasonal karstic Irish lakes). Journal of Plankton Research. 2011;**33**:385-403

Chapter 4

Environmental Factors and Cyanobacteria Genotype: Implications for Toxin Production

Maria Iasmina Moza, Ana Bianca Pavel and Carmen Postolache

Abstract

Starting from the well-known fact that a poor ecological status is associated with increasing phytoplankton abundance, greater proportions of cyanobacteria, and more frequent and intense toxic blooms, the main purpose of this chapter was to investigate abiotic and biotic factors influencing the ecology of Cyanobacteria populations and the expression of their toxigenic potential, by presenting results and conclusion both from field and laboratory studies found in the literature so far. A general overview of the complexity of factors is summarized at the end of the review in five scenarios about the Cyanobacteria behavior in freshwater ecosystems under different ecological statuses.

Keywords: cyanobacteria, freshwater, toxic bloom, mcyE, environmental factors, allelopathy

1. Introduction

In the last decades, toxic or so-called harmful algal blooms (HABs) have been increasingly reported on a global scale in frequency, distribution, and impact of their effects on public health, living resources, and local economies [1–3]. Such blooms occur not only in marine and estuarine environment, where there are about 150 harmful or toxic microalgal species [4], but also in the freshwater ecosystems impacted by eutrophication, where Cyanobacteria represent the dominant. Mass populations of toxic Cyanobacteria are a global phenomenon and the recent recognition that incidences of blooms may increase significantly under future climate change serves to reinforce further the seriousness of the potential risks to human health [1]. Due to this, toxic cyanobacteria have gained in recent years increasing amounts of attention by the scientists, authorities, and general public worldwide [5].

The cyanotoxins are secondary metabolites produced by about 40 species of Cyanobacteria [6], that include neurotoxic, hepatotoxic, genotoxic, inflammatory and cytotoxic agents [7]. The genera known for their potential ability to produce toxic substances include *Anabaena*, *Aphanizomenon*, *Cylindrospermopsis, Lyngbya, Microcystis, Nodularia*, *Nostoc,* and *Planktothrix* [8]; however, even if the species able to produce toxins are present, they may release toxins only in special circumstances, under the pressure of several anthropogenic or environmental factors. With over 90 known congeners globally, hepatotoxic microcystins (*mcyE* gene is involved in their synthesis) are among the most potent and commonly encountered toxins [9] and they are mainly produced by species belonging to the *Microcystis*, *Anabaena,* and *Planktothrix* genera, although many other taxa were shown to have toxic potential [10, 11]. Genetic-based methods enhanced the understanding of the natural distribution of genes that are involved in cyanotoxins production, despite only indicating potential toxin producers [12]. Quantitative real-time PCR (qPCR) has been increasingly applied to monitor potentially toxic cyanobacteria population shifts in diverse aquatic ecosystems worldwide [13, 14], due to its sensitivity and rapidity.

Although the ecology of Cyanobacteria is well described, the relationship between population dynamics and environmental factors that trigger the presence of genes involved in microcystins synthesis is poorly understood [15]. For example, the density of the hepatotoxic cells was directly correlated with the microcystins (MCs) concentration from water [16] and, in concordance, with the gene (*mcyE* for example) copies number inside the cells. Some strains may produce higher MC concentrations than others under the same laboratory conditions, while others can be more or less toxic depending on cultivating conditions [17]. In addition, the succession of Cyanobacteria species and biomass is influenced by seasonal changes in several factors including nutrients, grazing, light, and temperature, which affect also the concentration of MC in the field [18]. In environment, different strains produce different concentrations of MCs but not all the time target genes involved in MCs synthesis are well correlate with these concentrations as a recent study demonstrates [15]. The effects of environmental factors on the abundance of MC-producing and non-toxic *Microcystis* genotypes have, however, been studied on a limited scale [15].

A brief overview of the influence of several physic-chemical parameters is provided in [18] and summarized below: a positive correlation between toxic genotypes (*mcyA*) and nitrate (NO_3-N) concentrations was noticed, while temperature and orthophosphate (PO_4-P) concentration seem not to influence *mcyA* abundance [19]. In the same lake, a direct influence of the cyanophage assemblage in shifts of MC-producing and non-MC-producing subpopulations was suggested [20]. NH_4 and NO_3 were also shown to increase toxic (*mcyA*), *Microcystis* strains abundance, and MC concentrations in a hypereutrophic pond [21], without affecting the total *Microcystis* abundance. Total phosphorus was shown to positively correlate with both toxic (*mcyD*) and total *Microcystis* abundance in Lake Erie, USA, while *mcyD* genotypes correlated negatively with NO_3, total nitrogen, nitrogen to phosphorus ratio, and pH [22]. Phosphorus was also shown to correlate positively with the relative abundance of potentially hepatotoxic (*mcyE/ndaF*) cyanobacterial community of river Nile, Egypt, at a site showing strong phosphorus limitation [23].

In spite of the fact that both nitrogen and phosphorus influence the numbers/proportion of toxic genotypes, other factors such as low water depth, high pH, high temperature, lack of wind and water column stability, and nutrient and light availability favor cyanobacteria development [24] increasing the risk hazard.

Also, allelopathic relationships with macrophytes or other algal groups and grazing pressures of zooplankton or fish communities should be further considered as it has been noticed that competition and grazing stimulate toxin production [25, 26]. The cumulative impact of all these factors on toxin-producing cyanobacteria requires more detailed investigations to elaborate effective management measures to prevent health hazards. Within this review, we summarize the effect of biotic and abiotic factors on cyanobacteria abundance, genotype, and toxicity.

2. Interaction with environmental factors

Is well known that organisms interact with their living and non-living environment so, they are affected by *abiotic* (physical and chemical) and *biotic* (presence and activities of other organisms) factors. These adaptations are the result of evolution, the driving force of which is natural selection [27, 28]. Cyanobacteria play an invaluable role in freshwater ecosystems because of their ability to produce oxygen via oxygenic photosynthesis and convert atmospheric nitrogen to the biologically available form, ammonium (NH_4^+) [9, 29–31]. Additionally, their status as primary producers strengthens their connection to the aquatic environment, as they contribute to the foundation of the food web, especially stimulating bottom-up food web shifts in shallow, eutrophic lakes [32].

2.1 Abiotic factors

2.1.1 Physical and chemical parameters

Light conditions in shallow lakes may change on a time scale of days to weeks due to changes in cloudiness or wind-induced resuspension of sediments. It was demonstrated that light intensity is a critical factor influencing the production of cyanotoxins [33]. Changes in light conditions may profoundly affect the microcystin composition and thereby the toxicity of cyanobacteria [34]; the transcription of two genes responsible for microcystin production was already shown to be influenced by light quality [33]. For example, the harmful cyanobacterium *Planktothrix agardhii*, a species that prefers mostly shallow, turbid lakes, produces a more toxic variant during periods of sunny weather, when recreational activities in lakes are most attractive [34]. It has been documented that the excessive growth of Cyanobacteria can reduce water transparency with light penetration to only few centimeters, and thus having important effects on both pelagic and benthic communities [35]. The reduction of the euphotic zone together with the excessive increase in the ratio between the epilimnetic mixing layer and the euphotic depth is an unfavorable factor for other organisms [35].

Another physical parameter of the water, which has exhibited a high positive correlation with MCs concentration, but not with the number of *mcyE* gene copies, was the conductivity. Conductivity is a parameter related to the ability of electric conduction of water and can indicate the ion concentration. *Microcystis* utilizes various inorganic ions such as macronutrients and trace metals for growth [36]. This confirms previous studies showing that alkaline pH (7.5–8.5), electrical conductivity from 241 to 367 μS/cm, and temperature ranging from 24.8 to 32°C promote microcystin development [37].

The pH of water may also influence toxin production. For M. *aeruginosa*, higher MC production occurred at pH values above and below their optimum growth threshold [33].

Temperature is a crucial parameter of the environment that influences the survival, metabolism, growth, and reproduction of all Cyanobacteria, as well as the interactions between them and other species [38] as well as a fundamental relationship with cyanotoxin production [33]. For example, some cyanotoxins are often found at temperatures that would be considered sub-optimum for cell growth, with a maximum reported at 20°C and production ceasing at temperatures exceeding 35°C, namely cylindrospermopsin [33]. Anatoxin-a production has also been shown to be highest at 20°C, whereas maximum production of MC and nodularin has been reported to occur between 18 and 25°C [33]. Other studies showed that temperature, DO, and *Microcystis* biomass are positively correlated with MC accumulation as well [39]. Temperature alone may only partly determine bloom formation and it is accepted that a combination of factors are responsible for a bloom to develop: increasing temperatures, decreasing nutrients, and increased water column stability [40]; temperature has the most pronounced effect on toxicity; the highest toxicity was found at 20°C, but reduced at temperatures in excess of 28°C.

According to Paes and his colleagues [41], in all studied reservoirs that experienced toxic blooms, water transparency was reduced. *P. rubescens* abundance appears to be strongly influenced by water transparency, at least in pre-alpine lakes [42].

A series of mild winters with earlier ice break-up can lead to an earlier stratification and a shift in the composition of the phytoplankton from diatoms to Cyanobacteria [43]. After some scientists, the bloom of Cyanobacteria is associated with thermal stratification [44]. Other authors suggest that both growth and toxin production as *Anabaena* may be controlled by salinity [45].

2.1.2 Nutrients

Freshwater enrichment with nitrogen is a dynamic process, reflecting land use, hydrologic conditions, population, and economic growth [46, 47]. Recent field studies have shown that bloom production by *Microcystis* sp. (a widespread cyanobacterial genus frequently producing blooms) is often associated with high levels of nitrogen [48, 49]. Other experiments have demonstrated that toxic blooms are more likely to occur under elevated nitrogen concentrations [50].

Toxin production in non-nitrogen fixing Cyanobacteria, such as *Microcystis* and *Planktothrix*, has been shown to peak with high concentrations of nitrogen [33]. According to a recent study [51], total MC concentrations could be explained by a combination of abiotic factors like dissolved organic nitrogen (DIN), water temperature, and ammonium (NH_4^+). In this study, however, the best overall model to predict MC concentration combined both environmental variables and species biomass [51]. Total nitrogen, water temperature, ammonium, and dissolved organic nitrogen influenced the cyanobacterial community structure, which in turn resulted in differences in the dominant MC congener and the overall toxicity. Ammonium (NH_4^+) did not emerge as a significant variable in the multivariate model although it is considered important in structuring cyanobacterial communities based on empirical studies of some lakes [51].

The presence of N can contribute to both increased and decreased MC production, depending on the genera [33]. Nitrogen and phosphorus are the main elements for the matter and energy metabolism of the algae [52]. Toxin production in non-nitrogen fixing Cyanobacteria, such as *Microcystis* and *Planktothrix*, has been shown to peak with high concentrations of nitrogen [33]. When attempting biomanipulation, the omission of nitrogen causes approximately ten-fold decrease in toxicity [40]. High levels of nitrogen and phosphorus in freshwaters favor the growth of toxic strains over non-toxic ones [53]. Total nitrogen, pH, and the surface area of the lake predicted the occurrence probability of *mcyE* genes, whereas total phosphorus alone accounted for MC concentrations [54].

Historically, phosphorus was seen as the primary limiting nutrient in freshwaters [55], and now, it is generally accepted that reduction in phosphorus input is an effective way to inhibit cyanobacterial dominance in freshwater ecosystems. Unlike nitrogen, which can exist in gaseous form, phosphorus can only be found in dissolved ionic and particulate forms in natural waters. Cyanobacteria are able to dominate the phytoplankton communities at very low phosphorus concentrations because they have a very high natural affinity for this element [56, 57].

The findings related to phosphorous also demonstrate the importance of this nutrient to toxin production [33]. It is known that phosphorus was the limiting resource for phytoplankton biomass but the mixotroph *Dinobryon* (Chrysophyceae) biomass increased with decreasing total phosphorus concentrations [58]. There are studies where the regression analysis showed that the densities of phytoplankton and total phosphorus were positively and linearly correlated ($R = 0.715$) [52]. In the same study, the correlation between the concentration of TN and the density of phytoplankton was not remarkable ($R = 0.166$). P and N sources have been considered the main factors to affect the growth of *Microcystis* cells and MC levels, and such an influencing process is so complicated that different conclusions were obtained by different investigators from different lakes [59].

N:P ratio in water is an important parameter in controlling cyanobacterial blooms, with several data suggesting that total molar N:P ratios above 15 discourage the occurrence of these phenomena [60, 61]. After a screening of more than 240 water bodies, scientists concluded that maximum concentrations of microcystins occurred in hypereutrophic lakes at mass ratios of N:P below 23 and that many planktonic cyanobacteria have a benthic life stage where they engage in "luxury uptake" of P from sediments, and consequently, episodes of cyanobacterial recruitment from sediments can dramatically decrease the N:P ratio in the water column [62]. "TN/TP rule" hypothesis suggests that Cyanobacteria tended to dominate in the lake when TN/TP <29, while decreased when TN/TP >29, but this rule is less applicable to highly eutrophic waters when both N and P nutrients are very high [59]. It was also suggested that, at least in one highly eutrophic lake, TN/TP ratio is not the factor bursting of blooms of *Microcystis* [59]. *Microcystis* occurred only at higher TN, TP, temperature, and pH values [63], for example. The toxicity of *M. aeruginosa* is known to change depending on seasons [41, 64].

2.1.3 Variation in toxin concentration

Cyanotoxin production is dependent on a number of environmental conditions. Predominantly, these could include nutrient concentration, light intensity, and temperature [33]. Eutrophication increased the co-occurrence of potentially MC-producing cyanobacterial genera, raising the risk of toxic bloom formation [54]. In general, studies considering the effects of nutrients on toxin production find a positive correlation between the nutrient of interest and cellular toxin content. Research focusing on nitrogen, cyanobacterial growth, and subsequent toxin production report that increased nitrogen corresponds to increased toxin production [65–67]. Recent studies showed that physical and chemical parameters did not significantly account for both intracellular and dissolved microcystins occurrence [68].

World Health Organization has set a guideline value of 1 μg/l for microcystin-LR in drinking water [69]. It is notable that typical environmental concentrations of MC are below 10 μg/L [33].

For some organisms that accumulate microcystins, total MC concentrations declined after October and began to increase in May, from the season point of view, that give us also a clue about the toxin production in the ecosystem [39]. Microcystins production did not correlate with the high level of nutrients (*t* test, $p > 0.1$). In fact, microcystins were never detected in the more eutrophic reservoirs [68]. The cyanobacterial biomass in water, pH, and temperature also explains the variability of MC concentration in the sediment of shallow lakes [70]. MC and abundances of total *Microsytis* and MC-producing *Microsystis* (MCM) were shown to be positively correlated with pH, DO, and TP [59]. A significant difference was found for conductivity, phosphates, and the presence of microcystins Mc-LR and Mc-Tot in the oligo- and eutrophic reservoirs [68]. It is speculated that N and P nutrients and the associated genes (e.g., *mcy*) may jointly drive MC concentration and toxigenicity [59]. Also, species biomass was the best predictor of MC concentrations [51] and the highest growth rate is not correlated with the highest toxicity [40] that reveals that cell density may not be the best predictor for toxicity. *Microcystis* was often observed coexisting with *Anabaena, Planktothrix,* and *Phormidium* [59] and therefore can bloom together.

2.2 Biotic factors

2.2.1 Zooplankton

When the density of phytoplankton was too high, it was concluded that the multiplication of zooplankton was restrained or harmed by the worse eutrophication of the system [52]. Zooplankton and cyanobacteria may interact directly via the feeding and excretion of zooplankton [71] and indirectly via trophic linkage pathways and behavioral responses to physical and chemical parameters (light, temperature, turbulence, pH, and dissolved oxygen). Results of laboratory studies [72] suggest that many colonial cyanobacteria are either not eaten or are poor food for large zooplankton, particularly *Daphnia*. Therefore, at times when these colonial forms dominate the phytoplankton, *Daphnia* populations might be expected to show decreased growth and fecundity in response to food limitation or toxicity. It is very probably that any effect of cyanobacteria on zooplankton could be influenced by evolutionary mechanisms in natural systems with a long history of cyanobacterial blooms. Zooplankton that co-occur with dense biomass of Cyanobacteria have better chance to adapt than others that were not exposed to these conditions [41]. It was already demonstrated that, if the proportion of any toxic Cyanobacteria and any edible algae studied is in concordance with tested microorganism needs, the survival rate is growing [64, 73–76]. This suggests that in nature, at least some of the total amount of *M. aeruginosa* is consumed, but only when the strain is not highly toxic. Therefore, changes in the cyanobacterial abundances in nature are not only related to physical and chemical variables but are also likely due to grazing from zooplankton [64]. There were species-specific differences in the filtration and feeding rates of zooplankton when offered mixed diets of green algae and toxic cyanobacteria. These probably explain the coexistence of different zooplankton species in *Microcystis*-dominant waterbodies [76].

2.2.2 Rotifers

It has been previously shown that some rotifers like *Brachionus calyciflorus* react negatively to the presence of Cyanobacteria (*Microcystis aeruginosa* and *Synechococcus elongatus*) because they lack essential compounds and not because of their cell size [77]. On the other hand, when nutrient-rich green algae serve as food, the difference in the cell size matters and it is reflected directly in the lower growth rate of the rotifer population.

Among zooplankton, rotifers were shown to display a high range of tolerance to temperature. Thus, in some subtropical shallow lakes, the rotifer density increased with the increase in temperature, reaching its maximum at approximately 23°C but decreased slightly when the temperature exceeded 25°C [78]. An experiment focused on the effect of *Microcystis aeruginosa* on the survival and reproduction of the rotifer *Brachionus calyciflorus* using the life table method at different temperatures showed that *M. aeruginosa* suppressed the survival and reproduction of *B. calyciflorus*, particularly at a concentration of 10^6 cells/mL according to [79].

Some laboratory experiments have shown that temperature can play an important role in modifying the effect of toxic cyanobacteria on freshwater and planktonic herbivores. Thus, in rotifers acclimated for many generations to low (12–14°C), intermediate (19°C), and high (25–26°C) temperature, susceptibility to the cyanobacterium and its toxin increased significantly with temperature. The results indicate that seasonal increases in water temperature and climate warming may exacerbate the impact of toxic cyanobacteria on rotifers like *Brachionus calyciflorus* and *Asplanchna girodi*, and perhaps other zooplankton taxa [80]. So, it is clear that a potentially modifying effect of temperature needs to be considered when investigating the effect of the dynamics of zooplankton populations on toxic Cyanobacteria as well. For example, temperature is one of the important variables affecting the population growth and reproduction of rotifers including *B. calyciflorus*; higher temperatures (30°C) stimulate their population growth [81]. Usually, when two or more rotifer species compete for limited resources, one or more of them may be adversely affected by the presence of others [64], so bad food quality is not the only limiting factor for rotifers to be taken into account.

Even if the Cyanobacteria strain is non-toxic or toxic, at least in the case of *B. calyciflorus*, this type of diet suppresses the population growth; effects were less severe than those produced by toxic *M. aeruginosa* [73]. It was observed that in eutrophic systems, a higher abundance of rotifers is often observed with a higher abundance of Cyanobacteria, which indicates that not all cyanobacteria inhibit rotifer growth [41, 74]. Additionally, rotifers are affected by three selected forms (unicellular, filamentous, and colonial) of Cyanobacteria [73]. *Raphidiopsis raciborskii* (formerly *Cylindrospermopsis raciborskii*) is less harmful to the rotifer *Brachionus calicyflorus* than the worldwide occurring cyanobacterium *Microcystis aeruginosa*; *Microcystis* did not support the growth of *Brachionus*, but even killed them within the 2-day experimental period [74]. *Asplanchna girodi* was sensitive to the toxins of *Dolichospermum flos-aquae* (formerly *Anabaena flos-aquae*) and *Lyngbya majuscula* [82]. Moreover, the toxin-producing *Dolichospermum flos-aquae* has been shown to decrease the lifespan, fecundity, and population growth rate of the rotifers *Brachionus calycijlorus* and *Synchaeta pectinata* [33]. Usually, MCs are released into the water after demise of cells and dissolved MCs during the collapse of heavy blooms can come into contact with a wide range of aquatic organisms including rotifers and have adverse effects on them by increasing mortality [81]. Notably, different species in the ecosystem react differently to the toxins [64]. For example, the impact of *M. aeruginosa* in the diet had a more adverse effect on *B. calyciflorus* than on *B. havanaensis*, [64] but the same strain managed to kill the cladocerans *Ceriodaphnia dubia* and *Moina macrocopa* (Arthropoda) in the previous study [75].

Important to be mentioned is that rotifer density and water temperature were negatively correlated [83]. Depending on the ecosystem type (either temperate or tropical), rotifers react differently: some species could survive if they are exposed to *Raphidiopsis/Cylindrospermopsis*, but can be profoundly affected if they are exposed to *Microcystis,* for example [41]. However, when analyzed at the population level, rotifers and cladocerans showed a weak positive response to microcystin. For rotifers, negative relationships were observed for *Brachionus calyciflorus, Conochilus* sp., *Hexarthra* sp. with respect to the different orders, whereas calanoid copepods showed a positive response to microcystin [41].

Rotifers consume half of what cladocerans consume in terms of food; that could be an explanation why rotifers resist longer in nature if the ecosystem experiences a toxic bloom [76]. Also, in the same study, it was revealed that cladocerans' filtrate rate is growing if the food quality is lower, so again another motive because their survival rate is less: they manage to eat more *Microcystis* cells compared to rotifers in the same time interval.

Sometimes, a higher density of zooplankton, particularly rotifers, showed a worse trophic state of the ecosystem [52]. This suggests that, in some species, toxin production is likely linked to defense mechanisms against protozoan predation [33]. It has already been confirmed that toxins produced by *M. aeruginosa* did not change significantly the protozoan's mobility, morphology, or viability. In contrast, microcystins produced by *Gloeotrichia echinulata* were lethal for *Paramecium caudatum* (Ciliophora), while extracts of *L. majuscula* acted very fast to provoke the lysis of the protozoan *Tetrahymena pyriformis* [82]. Another study demonstrated that *Microcystis aeruginosa* was unable to produce MC in response to direct and indirect exposure to a protozoan grazer, this suggests that toxin production could be linked with defense against protozoan grazing only for some species [33].

A study conducted on the ciliate *Nassula* sp. isolated from a water body with no history of toxic blooms and fed with a toxic strain of *Planktothrix agardhii* for 8 months showed that this species can survive feeding exclusively on toxic cyanobacteria over an extended period of time, despite increasing MC concentrations [84]. Conversely, a number of protozoan grazers are known to actively feed and grow on toxic Cyanobacteria, and in some cases, *Microcystis aeruginosa* was unable to produce MC in response to direct and indirect exposure to a protozoan grazer [33]. The toxins are probably accumulated in the ciliates and only fractions of MC remain in the water. Therefore, water often contains a lower concentration of cyanotoxins compared to the amount found in zooplankton tissues [85]. For these reasons, this group is seen as a potential biotic agent to reduce toxin levels in freshwater bodies that typically experience cyanobacterial blooms. However, the ability to synthesize microcystin does not seem to offer toxic *Microcystis* populations a significant defense against grazing by co-occurring zooplankton communities [86].

2.2.3 Other zooplankton

Unlike with rotifers, the presence of cladocerans can trigger MC production in *Microcystis* sp. More exactly, in the presence of high zooplankton abundance (*Daphnia pulex* and *Brachionus calyciflorus*), at low cell density of *Microcystis* sp., the MC concentration was significantly higher as compared to controls [87]. Therefore, cladocerans react differently compared to other zooplankton groups, for example, *Keratella cochlearis* was superior in competition with *Daphia pulex* under toxic *Microcystis* [88]. One must consider that natural mesozooplankton were better grazers of both toxic and non-toxic strains of *Microcystis* than their cultured counterparts [86].

A study designed to test the development of tolerance in several zooplankton species to MC in a range of temperatures showed that the ability to utilize *Microcystis* improved at 30°C in species like *Moina macrocopa*, *Daphnia carinata*, and *Hexarthra mira*, but significantly decreased in case of rotifers [89]. It was also shown that low concentrations of edible algae favor small-sized cladocerans, while high concentrations favor large-sized cladocerans, such as *Daphnia*, but the presence of cyanobacteria can affect the dominance status of large-bodied daphnids, especially. Some cladocerans can coexist well with *Microcystis* sp. in nature but colony size affects cladoceran population and their interactions [88]. There are also strong intraspecific differences in the tolerance of different *Daphnia* clones to toxic/non-toxic cyanobacteria, and therefore, the dynamics of the daphnid populations vary significantly in the presence of these microalgae in their diet [90].

Zooplankton groups may act as vectors of the toxin uptake in the aquatic food web [91] and it seems that toxins are bioaccumulated in the ciliates and in the water, remain mostly fractions of microcystins (referring to their chemical structure) [84]. Therefore, water frequently contains a lower concentration of cyanotoxins compared to the amount found in zooplankton tissues [85].

2.2.4 Macrophytes and algae

It is well known that one of the more serious impacts of eutrophication on aquatic ecosystems is the disappearance of submerged macrophytes and the shift to a phytoplankton-dominated state [92]. When aquatic plants cover the wetland by 60% of the water surface, the equilibrium of the ecosystem was maintained and the proportion of the Cyanobacteria was maintained below 25%, almost the same with other algae groups namely Chlorophyta (36.8%) and Bacillariophyceae (31.0%) [52].

Experiments provide evidence that resource competition can occur between benthic and water column primary producers [93] and studies have shown that macrophytes can successfully suppress the growth of algae through releasing allelochemicals in nature and in experimental systems [94]. However, there is a lack of studies to integrate laboratory and field observations with respect to establishing allelopathic effects of macrophytes [33]. There are good hints for allelopathic mode of action of cyanobacterial secondary metabolites within a lake phytoplankton community but not much is known on the mechanisms of interactions between cyanobacteria and algae, and how both sides contribute to phytoplankton dynamics during the year [92].

Some charophytes germinated less in the presence of MC in the sediment, and they also had lower chlorophyll concentrations. Different species have displayed variable sensitivity to the presence of MC in the water [95]. MC-LR affects macrophyte *Ceratophyllum demersum* (loss of pigmentation, loss of leaves) [96]. High concentrations of exudates and extracts of *M. aeruginosa* can allelopathically inhibit both seed germination and the early growth and photosynthesis of *Potamogeton malaianus* seedlings [97]. A reduced growth in the presence of *M. aeruginosa* was observed for *Lemna minor* and for submerged plant *Ceratophyllum demersum* and also a significant decrease of chlorophyll a and b as well as total carotenoids [92]. MC-LR has also been shown to exert inhibitory effects on aquatic plants, such as *Ceratophyllum demersum*, with the toxin inhibiting growth, morphology, and photosynthesis at environmentally relevant concentrations (5 μgL^{-1}) [33] or 90% inhibition of photosynthesis *Elodea canadensis* [92].

There is one study that shows that golden algae have a great potential for biodegrading microcystin-LR (MC-LR) [98]. They reported that the alga *Poterioochromonas* sp. (Chrysophyceae) was able to degrade MC-LR in cells of *M. aeruginosa* while digesting the whole cells; the degradation process was determined to be carried out inside the algae cell. Also, another study showed that *Ochromonas* sp. (Chrysophyceae) was able to feed on all four cyanobacterial strains tested, including the very toxic single-celled strain PCC 7806 [63].

Notably, symbiosis between diatoms and cyanobacterial colonies may also occur in natural water ecosystems [99]. Cyanobacterial toxin production can be regulated by complex growth phase-dependent and environmental parameters and suppressed by the presence of extracellular products of a eukaryotic green alga like *Chlamydomonas reinhardtii* (Chlorophyta) [100].

However, Microcystin-LR extracted from *Microcystis aeruginosa* had a negative effect on the growth of several green algae [33].

3. Conclusions

In conclusion, since we cannot discuss one or a few direct factors that trigger the cyanobacteria mass development, target genes involved in cyanotoxin production and toxin production, and release into the environment, we formulate, in this short review chapter, five possible scenarios of cyanobacteria behavior in any freshwater ecosystem, especially shallow lakes (**Figures 1–5**).

Scenario 1 (**Figure 1**). Interpretation\explanation: in this ideal case, in our lake ecosystem, there are also fish, zooplankton (especially daphnia), phytoplankton, bacterioplankton, and aquatic plants, an ideal freshwater ecosystem. Cyanobacteria from here do not need to release toxins to any of other trophic compartment because: (i) there exists enough zooplankton that will eat microalgae, so those are not a stress factors for our Cyanobacteria; (ii) there exists enough fish that will eat the zooplankton so that is not a stress factor for our Cyanobacteria community; (iii) there are enough plants that will keep under control an algal bloom so that our zooplankton and fish could survive, but not that many that could disturb our Cyanobacteria community.

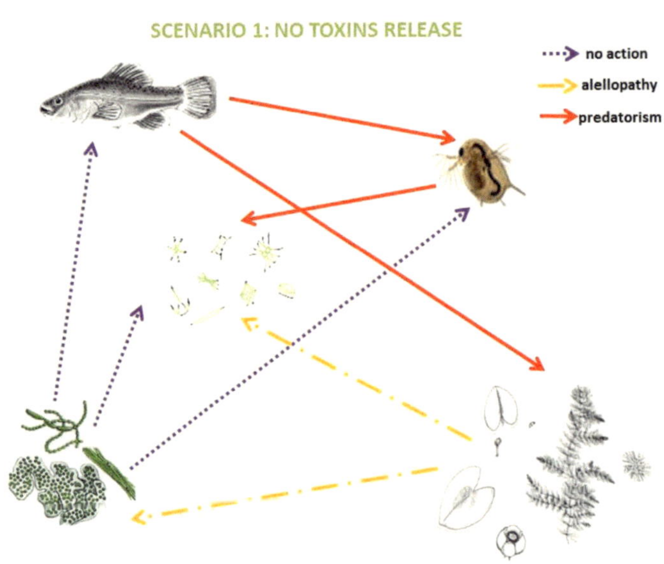

Figure 1.
First scenario of cyanobacteria behavior in a freshwater ecosystem: no toxin release.

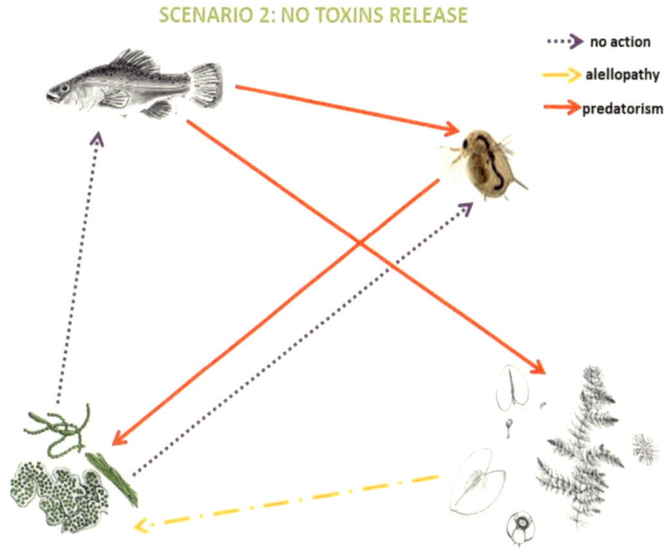

Figure 2.
Second scenario of cyanobacteria behavior in a freshwater ecosystem: no toxin release.

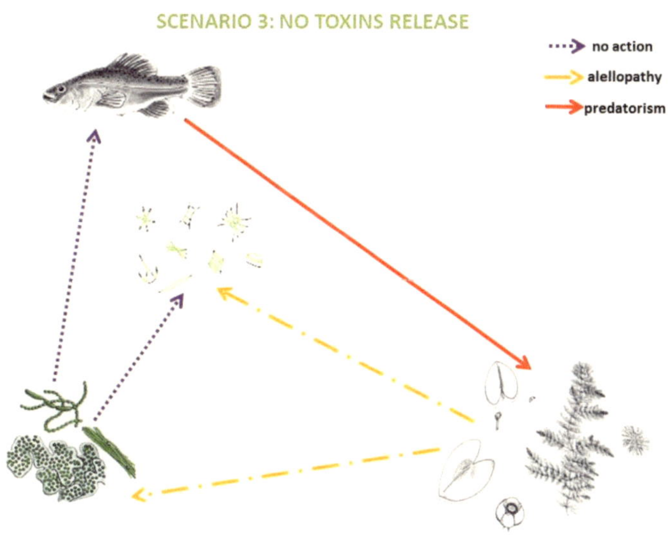

Figure 3.
Third scenario of cyanobacteria behavior in a freshwater ecosystem: no toxin release.

Conclusion: *this represents the ideal picture of a typical lake ecosystem and a scenario where Cyanobacteria could live in "no stress".* Lake type: oligo-mesotrophic.

Scenario 2 (**Figure 2**). Interpretation\explanation: in second case, in our lake ecosystem, we have also fish, zooplankton, (especially daphnia), bacterioplankton, and aquatic plants but less phytoplankton, because we have cyanobacteria blooming. Cyanobacteria from here do not need to release toxins to any of other trophic compartment because: (i) there exists enough zooplankton that will eat Cyanobacteria (because those are the main food source), but will not survive enough to become a

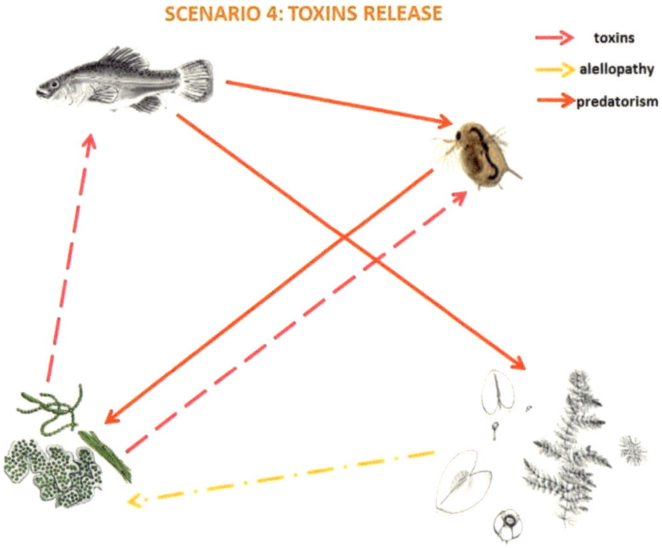

Figure 4.
Fourth scenario of cyanobacteria behavior in a freshwater ecosystem: toxin release.

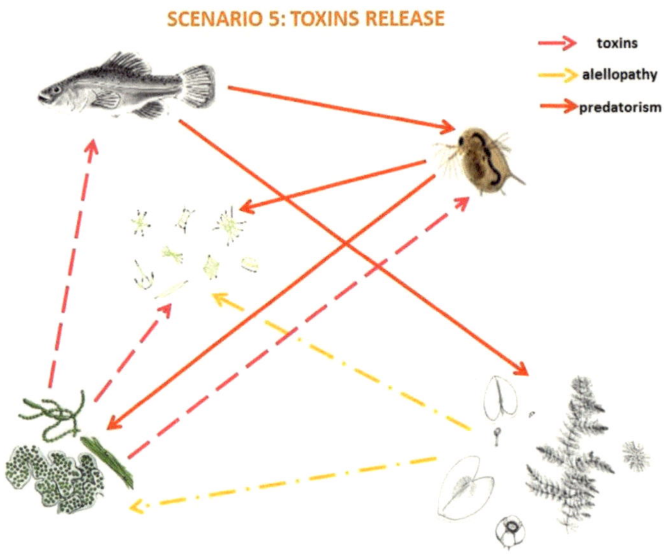

Figure 5.
Fifth scenario of cyanobacteria behavior in a freshwater ecosystem: toxin release.

stress factor for our Cyanobacteria population; (ii) there is enough fish that will eat plants and the zooplankton so that will be kept under control limit and not eat very much Cyanobacteria; (iii) there are some plants that will keep under control a massive algal bloom so that our zooplankton and fish could survive, but not that many that could disturb our cyanobacteria community. Conclusion: *this represents the picture of a small Cyanobacteria blooming but where the environmental changes are not that dramatic like in hypereutrophic lake, for example, so our Cyanobacteria have only "low stress".* Lake type: meso-eutrophic.

Scenario 3 (**Figure 3**). Interpretation\explanation: in this ideal case, in our lake ecosystem, there are also fish, phytoplankton, bacterioplankton, and aquatic plants, but no cladocerans, for example. Cyanobacteria from here do not need to release toxins to any of the other trophic compartments because: (i) there is enough fish that will eat other microalgae and zooplankton so that they do not represent as much of a stress factor for our Cyanobacteria community; (ii) there are enough plants that will keep under control an algal bloom by other microalgae so that our zooplankton and fish could survive, but not that many that could disturb our Cyanobacteria community. Conclusion: *this represents the ideal picture of a typical lake ecosystem and a scenario where Cyanobacteria could live in "medium stress".* Lake type: meso-eutrophic.

Scenario 4 (**Figure 4**). Interpretation\explanation: in this ideal case, in our lake ecosystem, there are also fish, zooplankton, bacterioplankton, and aquatic plants, but no other microalgae, because we have a Cyanobacteria bloom. Cyanobacteria from here began to release toxins to any of the other trophic compartments because: (i) there exists enough zooplankton that will eat them because they do not have other microalgae, so those are stress factors for our cyanobacteria; (ii) there do not exist enough fish that will eat the zooplankton and keep them under control; (iii) there are some plants that are struggling to survive, so they release allelopathic compounds against cyanobacteria, so this is another stress factor. Conclusion: *this represents the picture for instability of a typical lake ecosystem, if one important group is missing, and a scenario where Cyanobacteria live in "high stress" and become toxic.* Lake type: eu-hypertrophic.

Scenario 5 (**Figure 5**). Interpretation\explanation: in this ideal case, in our lake ecosystem, there are also fish, zooplankton, phytoplankton, bacterioplankton, and aquatic plants, but the diversity is low regarding species. We do not have Cyanobacteria bloom. Cyanobacteria from here began to release toxins to any of the other trophic compartments because: (i) there exists too much zooplankton that will eat microalgae, but also cyanobacteria, so those are stress factors for our cyanobacteria; (ii) it does not exist enough fish that will eat the zooplankton so that is a stress factor for our Cyanobacteria community; (iii) there are not enough fish to consume the zooplankton, which creates a stress factor for the cyanobacteria community. Conclusion: *this represents the ideal picture for the instability of a typical lake ecosystem and a scenario where Ccyanobacteria live in "high stress" and become toxic.* Lake type: eu-hypertrophic.

Acknowledgements

The authors would like to express their sincere gratitude to the Vice-Rector of Research at UB, Prof. Dr. Mariana Carmen Chifiriuc, for her invaluable guidance and support, as well as to the editor of this book, the reviewers, and the entire staff. We also acknowledge the financial support of the University of Bucharest for the chapter publication thought of the project CNFIS-FDI-2024-F-0484 INOVEX: Supporting the Development of New Competencies and Innovative Tools for Excellence in Interdisciplinary Research at the University of Bucharest for the. The authors would also like to acknowledge the previous projects that provided financial support during specific periods, contributing to this study. These include the Swiss Enlargement Contribution (IZERZ0 – 142165, 'CyanoArchive') under the Romanian-Swiss Research Programme; the Core Program within the National Research, Development, and Innovation Plan 2022–2027, supported by MRID (project no. 23020101, SIA-PRO, contract no. 7 N/2022); the 'Analysis of the potential for sustainable use of vegetation specific to the Danube-Danube Delta-Black Sea system – D3MN' (Code 108630), co-financed by the European Regional Development Fund; and the National Core Program—Romanian Ministry of Research and Innovation (project no. 25 N/2019, BIODIVERS 19270103). Special thanks are also extended to the experts from the Genetic Diversity Center, Zurich, for their valuable discussions, as well as to ICUB for their support.

Conflict of interest

The authors declare no conflict of interest.

Author details

Maria Iasmina Moza[1,2,3,4*], Ana Bianca Pavel[5] and Carmen Postolache[1]

1 Faculty of Biology, University of Bucharest, Bucharest, Romania

2 Research Institute of University of the Bucharest (ICUB), University of Bucharest, Bucharest, Romania

3 Moza Foundation, Bucharest, Romania

4 National Institute of Research and Development for Biological Sciences, Bucharest, Romania

5 National Research and Development Institute for Marine Geology and Geoecology, GeoEcoMar, Constanta, Romania

*Address all correspondence to: iasmina_moza@yahoo.com; maria-iasmina.moza@bio.unibuc.ro

References

[1] Paerl HW, Huisman J. Blooms like it hot. Science. 2008;**320**(5872):57-58

[2] Dolah FMV, Roelke D, Greene RM. Health and ecological impacts of harmful algal blooms: Risk assessment needs. Human and Ecological Risk Assessment: An International Journal. 2001;**7**(5):1329-1345

[3] Landsberg JH. The effects of harmful algal blooms on aquatic organisms. Reviews in Fisheries Science. 2002;**10**(2):113-390

[4] Landsberg JH et al. Saxitoxin puffer fish poisoning in the United States, with the first report of *Pyrodinium bahamense* as the putative toxin source. Environmental Health Perspectives. 2006;**114**(10):1502-1507

[5] Stewart I, Seawright AA, Shaw GR. Cyanobacterial poisoning in livestock, wild mammals and birds–an overview. In: Cyanobacterial Harmful Algal Blooms: State of the Science and Research Needs. New York, NY: Springer; 2008. pp. 613-637

[6] Westrick JA et al. A review of cyanobacteria and cyanotoxins removal/inactivation in drinking water treatment. Analytical and Bioanalytical Chemistry. 2010;**397**(5):1705-1714

[7] Tyler AN et al. Strategies for monitoring and managing mass populations of toxic cyanobacteria in recreational waters: A multi-interdisciplinary approach. Environmental Health. 2009;**8**(1):1-8

[8] Carmichael WW. Health effects of toxin-producing cyanobacteria: The CyanoHABs. Human and Ecological Risk Assessment: An International Journal. 2001;7(5):1393-1407

[9] Codd GA, Morrison LF, Metcalf JS. Cyanobacterial toxins: Risk management for health protection. Toxicology and Applied Pharmacology. 2005;**203**(3):264-272

[10] Ngwa FF, Madramootoo CA, Jabaji S. Comparison of cyanobacterial microcystin synthetase (mcy) E gene transcript levels, mcy E gene copies, and biomass as indicators of microcystin risk under laboratory and field conditions. MicrobiologyOpen. 2014;**3**(4):411-425

[11] Sivonen K. Emerging high throughput analyses of cyanobacterial toxins and toxic cyanobacteria. In: Cyanobacterial Harmful Algal Blooms: State of the Science and Research Needs. New York, NY: Springer; 2008. pp. 539-557

[12] Sivonen K, Börner T. Bioactive Compounds Produced by Cyanobacteria, in the Cyanobacteria. Poole, UK: Caister Academic Press; 2008. pp. 159-197

[13] Pearson LA, Neilan BA. The molecular genetics of cyanobacterial toxicity as a basis for monitoring water quality and public health risk. Current Opinion in Biotechnology. 2008;**19**(3):281-288

[14] Ostermaier V, Kurmayer R. Application of real-time PCR to estimate toxin production by the cyanobacterium *Planktothrix* sp. Applied and Environmental Microbiology. 2010;**76**(11):3495-3502

[15] Beversdorf LJ, Miller TR, McMahon KD. Long-term monitoring reveals carbon–nitrogen metabolism key to microcystin production in eutrophic lakes. Frontiers in Microbiology. 2015;**6**:456

[16] Sivonen K, Jones G. Cyanobacterial toxins. Toxic Cyanobacteria in Water: A Guide to Their Public Health Consequences, Monitoring and Management. 1999;**1**:43-112

[17] Vaitomaa J et al. Quantitative real-time PCR for determination of microcystin synthetase E copy numbers for microcystis and anabaena in lakes. Applied and Environmental Microbiology. 2003;**69**(12):7289-7297

[18] Martins A, Vasconcelos V. Use of qPCR for the study of hepatotoxic cyanobacteria population dynamics. Archives of Microbiology. 2011;**193**(9):615-627

[19] Yoshida M et al. Dynamics of microcystin-producing and non-microcystin-producing microcystis populations is correlated with nitrate concentration in a Japanese lake. FEMS Microbiology Letters. 2007;**266**(1):49-53

[20] Yoshida M et al. Ecological dynamics of the toxic bloom-forming cyanobacterium *Microcystis aeruginosa* and its cyanophages in freshwater. Applied and Environmental Microbiology. 2008;**74**(10):3269-3273

[21] Ha JH, Hidaka T, Tsuno H. Quantification of toxic Microcystis and evaluation of its dominance ratio in blooms using real-time PCR. Environmental Science & Technology. 2009;**43**(3):812-818

[22] Rinta-Kanto JM et al. Lake Erie *Microcystis*: Relationship between microcystin production, dynamics of genotypes and environmental parameters in a large lake. Harmful Algae. 2009;**8**(5):665-673

[23] Amer R, Díez B, El-Shehawy R. Diversity of hepatotoxic cyanobacteria in the Nile Delta, Egypt. Journal of Environmental Monitoring. 2009;**11**(1):126-133

[24] Kardinaal WEA, Visser PM. Dynamics of cyanobacterial toxins. In: Harmful Cyanobacteria, Aquatic Ecology Series. Dordrecht, Netherlands: Springer; 2005. pp. 41-63

[25] Gross EM, Feldbaum C, Graf A. Epiphyte biomass and elemental composition on submersed macrophytes in shallow eutrophic lakes. Hydrobiologia. 2003;**506**(1):559-565

[26] Legrand C et al. Allelopathy in phytoplankton-biochemical, ecological and evolutionary aspects. Phycologia. 2003;**42**(4):406-419

[27] Botnariuc N, Vădineanu A. Ecologie. Bucharest, Romania: Editura Didactică şi Pedagogică; 1982

[28] Darwin C. The Origin of Species and the Descent of Man. New York: The Modern Library; 1859

[29] Whitton BA, Potts M. Introduction to the cyanobacteria. In: Ecology of Cyanobacteria II. Dordrecht, Netherlands: Springer; 2012. pp. 1-13

[30] Huisman J, Hulot FD. Population dynamics of harmful cyanobacteria. In: Harmful Cyanobacteria. Dordrecht, Netherlands: Springer; 2005. pp. 143-176

[31] Visser S, Stroosnijder L, Chardon W. Nutrient losses by wind and water, measurements and modelling. Catena. 2005;**63**(1):1-22

[32] Muylaert K et al. Relationship between bacterial community composition and bottom-up versus top-down variables in four eutrophic shallow lakes. Applied and Environmental Microbiology. 2002;**68**(10):4740-4750

[33] Holland A, Kinnear S. Interpreting the possible ecological role(s) of cyanotoxins: Compounds for competitive advantage and/or physiological aide? Marine Drugs. 2013;**11**(7):2239-2258

[34] Tonk L et al. The microcystin composition of the cyanobacterium *Planktothrix agardhii* changes toward a more toxic variant with increasing light intensity. Applied and Environmental Microbiology. 2005;**71**(9):5177-5181

[35] Sukenik A, Quesada A, Salmaso N. Global expansion of toxic and non-toxic cyanobacteria: Effect on ecosystem functioning. Biodiversity and Conservation. 2015;**24**(4):889-908

[36] Tao M et al. Use of a generalized additive model to investigate key abiotic factors affecting microcystin cellular quotas in heavy bloom areas of Lake Taihu. PLoS One. 2012;7(2):10

[37] Carvalho LRD et al. A toxic cyanobacterial bloom in an urban coastal lake, Rio Grande do Sul state, southern Brazil. Brazilian Journal of Microbiology. 2008;**39**(4):761-769

[38] Ibelings BW et al. Chytrid infections and diatom spring blooms: Paradoxical effects of climate warming on fungal epidemics in lakes. Freshwater Biology. 2011;**56**(4):754-766

[39] Xue Q et al. Temporal dynamics of microcystins in *Limnodrilus hoffmeisteri*, a dominant oligochaete of hypereutrophic Lake Taihu, China. Environmental Pollution. 2016;**213**:585-593

[40] Oberholster P, Botha A-M, Grobbelaar J. *Microcystis aeruginosa*: Source of toxic microcystins in drinking water. African Journal of Biotechnology. 2004;**3**(3):10

[41] Paes TA et al. Can microcystins affect zooplankton structure community in tropical eutrophic reservoirs? Brazilian Journal of Biology. 2016;**76**(1):202-209

[42] Ernst B et al. Abundance and toxicity of *Planktothrix rubescens* in the pre-alpine Lake Ammersee, Germany. Harmful Algae. 2009;**8**(2):14

[43] Adrian R et al. Effects of ice duration on plankton succession during spring in a shallow polymictic lake. Freshwater Biology. 1999;**41**(3):621-634

[44] Dantas ÊW, Moura AN, Bittencourt-Oliveira MdC. Cyanobacterial blooms in stratified and destratified eutrophic reservoirs in semi-arid region of Brazil. Anais da Academia Brasileira de Ciências. 2011;**83**(4):1327-1338

[45] Engström-Öst J, Repka S, Mikkonen M. Interactions between plankton and cyanobacterium *anabaena* with focus on salinity, growth and toxin production. Harmful Algae. 2011;**10**(5):530-535

[46] Martínez-Espinosa C et al. Denitrification in wetlands: A review towards a quantification at global scale. Science of the Total Environment. 2021;**754**:142398

[47] Galloway JN et al. The nitrogen cascade. Bioscience. 2003;**53**(4):341-356

[48] Ma H et al. Growth inhibitory effect of microcystis on *Aphanizomenon flos-aquae* isolated from cyanobacteria bloom in Lake Dianchi, China. Harmful Algae. 2015;**42**:43-51

[49] Davis TW et al. Effects of nitrogenous compounds and phosphorus on the growth of toxic and non-toxic strains of microcystis during cyanobacterial blooms. Aquatic Microbial Ecology. 2010;**61**(2):149-162

[50] Vézie C et al. Effect of nitrogen and phosphorus on growth of toxic and

nontoxic Microcystis strains and on intracellular microcystin concentrations. Microbial Ecology. 2002;**43**(4):443-454

[51] Monchamp ME et al. Nitrogen forms influence microcystin concentration and composition via changes in cyanobacterial community structure. PLoS One. 2014;**9**(1):e85573

[52] Cao L et al. Biodiversity and water quality variations in constructed wetland of Yongding River system. Acta Ecologica Sinica. 2007;**27**(9):3670-3677

[53] Vezie C et al. Effect of nitrogen and phosphorus on growth of toxic and nontoxic Microcystis strains and on intracellular microcystin concentrations. Microbial Ecology. 2002;**43**(4):443-454

[54] Rantala A et al. Detection of microcystin-producing cyanobacteria in Finnish lakes with genus-specific microcystin synthetase gene E (mcyE) PCR and associations with environmental factors. Applied and Environmental Microbiology. 2006;**72**(9):10

[55] Paerl H. Nutrient and other environmental controls of harmful cyanobacterial blooms along the freshwater–marine continuum. In: Cyanobacterial Harmful Algal Blooms: State of the Science and Research Needs. New York, NY: Springer; 2008. pp. 217-237

[56] Posselt AJ, Burford MA, Shaw G. Pulses of phosphate promote dominance of the toxic cyanophyte *Cylindrospermopsis raciborskii* in a subtropical water reservoir 1. Journal of Phycology. 2009;**45**(3):540-546

[57] Dolman AM et al. Cyanobacteria and cyanotoxins: The influence of nitrogen versus phosphorus. PLoS One. 2012;**7**(6):e38757-e38757

[58] Kamjunke N, Henrichs T, Gaedke U. Phosphorus gain by bacterivory promotes the mixotrophic flagellate *Dinobryon* spp. during re-oligotrophication. Journal of Plankton Research. 2006;**29**(1):39-46

[59] Yu G et al. Variation of Microcystis and microcystins coupling nitrogen and phosphorus nutrients in Lake Erhai, a drinking-water source in southwest plateau, China. Environmental Science and Pollution Research International. 2014;**21**(16):9887-9898

[60] Paerl HW et al. Controlling cyanobacterial blooms in hypertrophic Lake Taihu, China: Will nitrogen reductions cause replacement of non-N_2 fixing by N_2 fixing taxa? PLoS One. 2014;**9**(11):e113123

[61] Smith VH, Schindler DW. Eutrophication science: Where do we go from here? Trends in Ecology & Evolution. 2009;**24**(4):201-207

[62] Orihel DM et al. High microcystin concentrations occur only at low nitrogen-to-phosphorus ratios in nutrient-rich Canadian lakes. Canadian Journal of Fisheries and Aquatic Sciences. 2012;**69**(9):1457-1462

[63] Van Donk E et al. The effect of a mixotrophic chrysophyte on toxic and colony-forming cyanobacteria. Freshwater Biology. 2009;**54**(9):1843-1855

[64] Alva-Martínez AF et al. Effect of mixed toxic diets (Microcystis and chlorella) on the rotifers *Brachionus calyciflorus* and *Brachionus havanaensis* cultured alone and together. Limnologica—Ecology and Management of Inland Waters. 2009;**39**(4):302-305

[65] Gobler C et al. Interactive influences of nutrient loading, zooplankton grazing, and microcystin synthetase gene expression on cyanobacterial bloom dynamics in a eutrophic New York lake. Harmful Algae. 2007;**6**(1):119-133

[66] Downing TG et al. Medium N: P ratios and specific growth rate comodulatemicrocystin and protein content in *Microcystis aeruginosa* PCC7806 and *M. Aeruginosa* UV027. Microbial Ecology. 2005;**49**(3):468-473

[67] Ahn C-Y, Chung A-S, Oh H-M. Rainfall, phycocyanin, and N: P ratios related to cyanobacterial blooms in a Korean large reservoir. Hydrobiologia. 2002;**474**(1):117-124

[68] Asencio AD. Determination of microcystins in reservoirs of different basins in a semiarid area. Journal of Applied Phycology. 2013;**25**(6):10

[69] Chen X et al. Fates of *Microcystis aeruginosa* cells and associated microcystins in sediment and the effect of coagulation process on them. Toxins (Basel). 2014;**6**(1):152-167

[70] Song H et al. The importance of lake sediments as a pathway for microcystin dynamics in shallow eutrophic lakes. Toxins (Basel). 2015;7(3):900-918

[71] Haney JF. Field studies on zooplankton-cyanobacteria interactions. New Zealand Journal of Marine and Freshwater Research. 1987;**21**(3):467-475

[72] Lampert W. Laboratory studies on zooplankton-cyanobacteria interactions. New Zealand Journal of Marine and Freshwater Research. 1987;**21**(3):483-490

[73] Zhao S, Wang Y, Li D. Effects of toxic and non-toxic *Microcystis aeruginosa* in different mixtures with *Scenedesmus obliquus* on growth of *Brachionus calyciflorus*. Journal of Freshwater Ecology. 2014;**29**(3):377-386

[74] Soares MCS, Lurling M, Huszar VLM. Responses of the rotifer *Brachionus calyciflorus* to two tropical toxic cyanobacteria (*Cylindrospermopsis raciborskii* and *Microcystis aeruginosa*) in pure and mixed diets with green algae. Journal of Plankton Research. 2010;**32**(7):999-1008

[75] Alva-Martínez AF, Sarma SSS, Nandini S. Population dynamics of *Brachionus calyciflorus* and Brachionus havanaensis (Rotifera) on mixed diets with *Microcystis aeruginosa* and green algae. Hidrobiológica. 2007;**7**:10

[76] Pérez-Morales A, Sarma SS, Nandini S. Feeding and filtration rates of zooplankton (rotifers and clodocerans) fed toxic cyanobacterium (*Microcystis aeruginosa*). Journal of Environmental Biology. 2014;**35**(6):8

[77] Lürling M, Beekman W. Influence of food-type on the population growth rate of the rotifer *Brachionus calyciflorus* in short-chronic assays. Acta Ecologica Sinica. 2006;**51**(1):9

[78] Ji G, Wang X, Wang L. Planktonic rotifers in a subtropical shallow lake: Succession, relationship to environmental factors, and use as bioindicators. ScientificWorldJournal. 2013;**2013**:702942

[79] Zhang X, Geng H. Effect of *Microcystis aeruginosa* on the rotifer *Brachionus calyciflorus* at different temperatures. Bulletin of Environmental Contamination and Toxicology. 2012;**88**(1):20-24

[80] Gilbert JJ. Effect of temperature on the response of planktonic rotifers to a toxic cyanobacterium. Ecology. 1996;**77**(4):8

[81] Huang L et al. Responses in population growth and reproduction of the freshwater rotifer *Brachionus calyciflorus* to microcystin-LR at different temperatures. Annales de Limnologie -

International Journal of Limnology. 2012;**48**(4):383-390

[82] Vasconcelos V. Cyanobacteria toxins: Diversity and ecological effects. Limnetica. 2001;**20**(1):14

[83] Sellami I et al. Abundance and biomass of rotifers in relation to the environmental factors in geothermal waters in southern Tunisia. Journal of Thermal Biology. 2009;**34**(6):267-275

[84] Combes A et al. Ciliate *Nassula* sp. grazing on a microcystin-producing cyanobacterium (*Planktothrix agardhii*): Impact on cell growth and in the microcystin fractions. Aquatic Toxicology. 2013;**126**:435-441

[85] Ferrão-Filho AS, Kozlowsky-Suzuki B, Azevedo SM. Accumulation of microcystins by a tropical zooplankton community. Aquatic Toxicology. 2002;**59**:8

[86] Davis TW, Gobler CJ. Grazing by mesozooplankton and microzooplankton on toxic and non-toxic strains of Microcystis in the Transquaking River, a tributary of Chesapeake Bay. Journal of Plankton Research. 2010;**33**(3):415-430

[87] Pérez-Morales A, Sarma SSS, Nandini S. Microcystins production in Microcystis induced by *Daphnia pulex* (Cladocera) and *Brachionus calyciflorus* (Rotifera). Hidrobiológica. 2015;**25**(3):6

[88] Chen F, Xie P, Qin B. Different competitive outcomes among four species of cladocerans under different alga combinations of colonial *Microcystis* spp. and green alga *Scenedesmus obliquus*. Hydrobiologia. 2007;**581**(1):209-215

[89] Nandini S. Responses of rotifers and cladocerans to *Microcystis aeruginosa* (Cyanophyceae): A demographic study. Aquatic Ecology. 2000;**34**:16

[90] Drugă B et al. Cyanobacteria affect fitness and genetic structure of experimental daphnia populations. Environmental Science & Technology. 2016;**50**(7):3416-3424

[91] Smith JL, Haney JF. Foodweb transfer, accumulation, and depuration of microcystins, a cyanobacterial toxin, in pumpkinseed sunfish (*Lepomis gibbosus*). Toxicon. 2006;**48**(5):580-589

[92] Wiegand C, Pflugmacher S. Ecotoxicological effects of selected cyanobacterial secondary metabolites: A short review. Toxicology and Applied Pharmacology. 2005;**203**(3):201-218

[93] Cristofor S et al. Long-term changes of submerged macrophytes in the lower Danube wetland system. Hydrobiologia. 2003;**506**(1):625-634

[94] Gross EM et al. Searching for allelopathic effects of submerged macrophytes on phytoplankton—State of the art and open questions. Shallow Lakes in a Changing World. Springer; 2007. pp. 77-88

[95] Rojo C et al. Allelopathic effects of microcystin-LR on the germination, growth and metabolism of five charophyte species and a submerged angiosperm. Aquatic Toxicology. 2013;**144**:1-10

[96] Pflugmacher S. Promotion of oxidative stress in the aquatic macrophyte *Ceratophyllum demersum* during biotransformation of the cyanobacterial toxin microcystin-LR. Aquatic Toxicology. 2004;**70**(3):169-178

[97] Zheng G et al. Cyanobacteria can allelopathically inhibit submerged macrophytes: Effects of *Microcystis aeruginosa* extracts and exudates on *Potamogeton malaianus*. Aquatic Botany. 2013;**109**:1-7

[98] Zhang X et al. Isolation of a *Poterioochromonas* capable of feeding on *Microcystis aeruginosa* and degrading microcystin-LR. FEMS Microbiology Letters. 2008;**288**(2):241-246

[99] Snoeijs P, Murasi LW. Symbiosis between diatoms and cyanobacterial colonies. VIE MILIEU. 2004;**54**(2-3):8

[100] Kearns KD, Hunter MD. Green algal extracellular products regulate antialgal toxin production in a cyanobacterium. Environmental Microbiology. 2000;**2**(3):7

Chapter 5

Physicochemical Parameters' Effects on the Freshwater Cyanobacterium *Microcystis* Sp. and Their Toxins

Hichem Nasri, Loubna Ghodbane, Amina Amrani, Chaima Aounallah, Fateh Sedrati, Maaref Hanen, Zineb Labidi, Zakia Hammoum, Nassima Beldjoudi and Hadjer Zaidi

Abstract

Microcystis sp. are ubiquitous and highly prevalent Cyanobacteria found in freshwater ecosystems. They are known for episodic, large-scale proliferations known as "blooms", which are driven by factors such as elevated nitrogen and phosphorus concentrations, enhanced water column stability, and increased temperatures. *Microcystis* sp. are infamous for producing a class of potent hepatotoxins known as "microcystins (MCs)", which exert their toxicity primarily through the inhibition of serine/threonine protein phosphatases 1 (PP1) and 2A (PP2A). This inhibition disrupts cellular signal transduction pathways and affects numerous cellular processes by preventing the dephosphorylation of proteins. As a result, there is a significant disruption of the cytoskeleton, altered cellular signaling, and, ultimately, cell apoptosis. Additionally, microcystins interfere with cellular antioxidant defense mechanisms, causing oxidative stress by generating reactive oxygen species (ROS). This oxidative stress can lead to damage across various organs and tissues, including the liver, which is particularly susceptible to microcystin toxicity. This chapter provides a comprehensive overview of how physicochemical parameters influence the proliferation of *Microcystis* sp. and the production of microcystins. It explores the intricate relationship between environmental factors and the enhanced biosynthesis of these toxins, thereby elucidating the conditions that lead to their intensified production.

Keywords: *Microcystis* sp., cyanobacteria, microcystins, physiochemical parameters, prevention

1. Introduction

As primary producers, microalgae play a key role in nutrient cycling and are essential for maintaining the equilibrium of food web dynamics, maintaining energy flow, ecological balance, nitrogen fixation, material circulation, and pollution degradation [1, 2]. But they can also contribute to the accumulation of contaminants that can negatively impact the ecosystem [3, 4]. Cyanobacteria require specific conditions to perform their physiological processes effectively [5–7]. Cyanobacteria have been reported to contaminate water bodies such as lakes, lagoons, oceans, rivers, streams, reservoirs, and wells [8–11].

In the presence of certain environmental conditions, i.e., higher water temperatures, high light intensity, and eutrophication (excess nitrogen and/or phosphorous), the growth and dominance of algae species are favored, which can result in the formation of Blooms [12–14]. It has emerged as a global ecological and environmental problem [15, 16]. Excessive proliferation of cyanobacteria can lead either to hypoxia, by depleting dissolved oxygen levels, or the production of a variety of cyanotoxins [17], such as microcystins (MC), a group of over 248 congeners [18], the most toxic of which are MC-LR (leucine, arginine), MC-RR (arginine, arginine), and MC-YR (tyrosine, arginine) [18]. They can also generate a significant amount of odor compounds derived from algal volatile organic compounds (VOCs). These compounds severely impair the esthetic quality of water, substantially increase water treatment costs, prompt numerous consumer complaints about unpleasant odors, and can even precipitate drinking water crises [19, 20].

The genus *Microcystis* is one of the most important and major bloom-forming Cyanobacteria due to its ecological and public health importance [21, 22]. Microcystis, a cyanobacterium with a widespread distribution, often dominates phytoplankton communities in nutrient-rich drinking water sources. It produces gas vesicles that reduce its density relative to water. In stable water columns, buoyant *Microcystis* rises to the surface, creating dense blooms. Minaglia et al. [23] and Medrano et al. [24], in natural conditions, *Microcystis* occurs in colonies with sheaths and mucilage that protect the cells from zooplankton grazing, viral or bacterial attacks, desiccation, and other potential environmental threats, while also increasing their buoyancy [25–29]. They present a formidable challenge in freshwater ecosystems due to their pervasive distribution, frequent blooms, and substantial toxicity [16, 30]. It is well-established that toxic *Microcystis* blooms pose significant ecological hazards, with the potential to severely impact water quality, biodiversity, and human health [31, 32] and can induce oxidative stress in aquatic species [18, 33].

Numerous researchers have determined that both biotic factors (e.g., zooplankton, bacteria) and abiotic factors (e.g., light intensity, temperature) contribute to the colony formation of *Microcystis* [34–40]. The presence and synthesis of *Microcystis* toxins are significantly affected by environmental conditions such as elevated water temperatures, severe nitrogen-phosphorus imbalance (N), reduced salinity, and increased water clarity. Nevertheless, these influences remain contentious due to variations in geographic locations and seasonal changes [4, 41]. They are likely to be the first to benefit from the changing global climate [42, 43]. This proliferation has led to significant environmental, economic, and societal challenges, resulting in long-term negative impacts on water quality, fisheries, esthetics, tourism, and other economic activities [4].

This chapter summarizes the effects of physicochemical parameters on the potentially toxic species *Microcystis* sp., elucidating the relationship between these factors and the intensive production of toxins.

2. Effects of environmental parameters on *Microcystis sp.* cells

2.1 Effects of temperature

The role of temperature in cyanobacterial growth and bloom formation is well-established. It is considered one of the major determinants influencing nutrient uptake, phytoplankton growth, and spatial-temporal distribution in freshwater systems. The importance of this environmental factor for cyanobacterial growth and bloom formation in shallow water bodies is well documented.

2.1.1 Temperature positive effect

Table 1 presents the major influence of T°C fluctuation on *Microcystis sp.* Bloom.

2.1.2 Temperature negative effect

Table 2 resumes the interaction between T° and *Microcystis* sp. Bloom.

Both tables indicate that temperature fluctuations significantly impact the growth, metabolism, and toxin production of *Microcystis* sp. Cyanobacteria like *Microcystis* typically thrive within a specific temperature range, and deviations from this range can affect their physiological processes. Higher temperatures generally accelerate metabolic rates and cellular division, leading to enhanced algal growth and increased biomass accumulation. However, extreme temperature fluctuations, including both abrupt warming and cooling, can stress the organisms and disrupt their growth. Elevated temperatures often enhance the synthesis of microcystins, the potent hepatotoxins produced by Microcystis, by increasing the activity of the enzymes involved in their biosynthesis. Conversely, sudden cooling can slow down metabolic processes and reduce toxin production, but it may also lead to the release of toxins from lysed cells as the algae experience stress. Additionally, temperature-induced stratification of water bodies can create stable conditions favorable for *Microcystis* blooms, as warmer surface waters are less likely to mix with cooler, nutrient-rich deeper waters. Understanding how temperature fluctuations affect *Microcystis* is crucial for predicting bloom dynamics and managing the associated risks in freshwater systems.

2.2 Effects of pH

The pH of the water environment plays a crucial role in the growth and toxicity of *Microcystis* (**Table 3**). At higher pH levels, typically above 8, *Microcystis* tends to proliferate more rapidly, as alkaline conditions favor its photosynthetic processes. This increase in growth is often accompanied by elevated production of microcystins, the potent toxins produced by these cyanobacteria. Conversely, at lower pH levels, the growth of *Microcystis* is generally inhibited, leading to a reduction in toxin production. However, extreme pH levels, whether acidic or alkaline, can cause cell lysis, potentially releasing accumulated toxins into the water, thereby posing a risk to aquatic life and human health. Therefore, managing pH levels in water bodies is vital for controlling the spread and toxicity of *Microcystis* blooms. The variability of the pH compensation point (pHc) between strains of the cyanobacterium *Microcystis*

	T °C	Major Effect	Ref.
M. aeruginosa	28.80–30.50°C	• *M. aeruginosa* rapid growth.	[44]
M. aeruginosa	32.00°C	• A faster *M. aeruginosa* growth	[45]
Microcystin	24°C and 28.5°C	• Highest MC concentrations record (Kucukcekmece Lagoon)	[46]
M. aeruginosa FACHB915	−16°C and 32°C −24°C	• Increase in the biovolumes • Significant competitive advantage	[47]
M. aeruginosa FACHB469	1. 16°C, 24°C, 32°C. 2. at 24°C 3. at 32°C	• 1-Coexistence • 2- A higher growth rate + significant increase in total biovolume • 3-A slight increase in the growth rate	
M. aeruginosa FACHB905	−16°C, 24°C, 32°C.	• Dominance in the mixed population	
M. aeruginosa	- 24°C	• Increase in the growth rates • Advantages to outcompete other phytoplankton	[48]
M. aeruginosa	−20°C	• Perfect for growth and dominance	[49]
Microcystin	25°C or greater	• Pmax (photosynthetic capacity), Rest (specific respiration rate), and growth rate	[50]
M. aeruginosa complex (MAC)	-at 30°C	• Lowest values of colonies variation and cell-free space percentage	[51]
M. viridis	−25°C	Optimal activity of photosynthesis	[52]
M. aeruginosa	−30°C		
M. wesenbergii	−30°C		
Microcystis	During summer	*Microcystis* dominance with a biomass and ratio driven by T °C	[53]
MC-LR	−15°C	Slight *M. aeruginosa* growth	[54]
	at 15°C, 20°C, 25°C, and 30°C	An increase in intracellular MC-LR content	
	30°C	• The highest content of extracellular MC-LR	
	15 to 30°C	The capacity of cyanobacteria to produce MC-LR toxin was optimized	
Microcystis	Autumn and summer seasons	Water temperature was the main variable Influencing Cyanobacteria, *Microcystis*, and microcystin occurrence • *Microcystis* dominance	[55]
Microcystis	June and August	• Cyanobacterial density increase	[43]
Microcystin	20–25°C	• Temperature high degrees increase the chance of MC occurrence with the highest probability of having microcystin concentrations	[56]
Microcystis	summer season	• Climatic conditions such as temperature have the greatest impact on the biovolume of *Microcystis* • With higher temperatures biovolume would be higher	[57]
Microcystis	18–20°C	• Great amount of recruitment from culturing sediment samples	[58]

Table 1.
Effects of temperature on Microcystis *sp. growth.*

	T °C	Major Effect	Ref
M. aeruginosa	−32.50°C	A rapid decrease of the Bloom	[44]
M. aeruginosa	under 16.5°C	• *M. aeruginosa* was not recorded	[46]
Microcystis	-below 15°C	• *Microcystis* was the most severely limited	[50]
Microcystis	- below 14°C	• No cyanobacteria recruitment occurred.	[58]
M. viridis; M. aeruginosa; M. wesenbergii	−33 to 36°C	• Sharp decline in growth rates and photosynthetic activity	[59]
Microcystis	decrease of water temperature	• Gas vesicle formation reduction and hydrocarbon accumulation • A loss in *Microcystis* colonies buoyancy and make them sink to the bottom	[60]
M. aeruginosa	decrease of water temperature	• An inhibition in *M. aeruginosa* cell growth. • Increase in ROS levels • Reduction in the content of photosynthetic pigment and (PSII) photosystem II performance, • A gradual increase in 1O2 levels • β-carotene content decreased by quenching 1O2 with β- cyclocitral emission increasing • Negative effect on their development and proliferation	[61]
MC-LR	- at 35°C	• The content of intracellular MC-LR decreased • The capacity of Cyanobacteria to produce MC-LR toxin diminished	[54]

Table 2.
Negative effects of temperature on Microcystis *sp. Growth.*

aeruginosa was investigated by [73]. It was demonstrated that the pHc can be used to distinguish whether the organism is able to take up HCO_3 as an inorganic carbon (Ci) source in photosynthesis or not.

2.3 Light effect

Microcystis has a strong relationship with light, as it relies on sunlight for photosynthesis, which fuels its growth and proliferation. Adequate light exposure is essential for *Microcystis* to produce energy and, consequently, to thrive in aquatic environments. High light intensity often leads to increased photosynthetic activity, promoting the growth of *Microcystis* blooms. However, intense light can also induce stress, potentially leading to the production of protective pigments and even microcystins, the toxins associated with these Cyanobacteria. Conversely, low light conditions can limit the growth of *Microcystis*, reducing the likelihood of bloom formation and toxin production. Thus, light availability is a key factor in the dynamics of *Microcystis* populations and their associated risks. **Table 4** shows the effect of light changes on *Microcystis* strains' growth and metabolism.

Hesse et al. [82] conducted a study focusing on the comparison of growth and pigment content of strain PCC 7806 and its mcyB3 mutant deficient in microcystin biosynthesis, under semicontinuous culture conditions. Both wild-type and mutant

	Acidic	Neutral 6.5 to 7.5	Alkaline	ref
MC-LR	pH 1 + 40°C • MC-LR decomposition • Hydrolysis of the Mdha moiety to several linearized peptides. • The half-life of MC-LR was 3 weeks. • Adding methanol esterification predominated over hydrolysis	/	Ph 9 + 21 30°C • MC-LR decomposition • Hydrolysis of the Mdha moiety to several linearized peptides. • The half-life of MC-LR was 10 weeks	[62]
MC -RR	/	Ph 7.0: • An increase in MC-RR productivity	Ph 9.2: • An increase in MC-RR productivity	[63]
MC-LR	pH = 1: • Log Dow of MC-LR was 2.18	/	• pH = 10: • Log Dow of MC-LR decreased to −1.76. • The Cyanobacteria may be flourishing in a basic pH environment	[64]
MC-LR	pH 1.0: • Log DOW was 1.63	pH 6.5: • log DOW decreased to 1.26.	pH of 6.5 ~ 12.0: • Log DOW stabilized between −1.04 and − 1.56.	[65]
MC-RR	pH of 1.00 ~ 4.00: • Log DOW varied between −1.24 and − 0.67	/	pH of 4.00 ~ 12.00: • Log DOW stabilized between −1.20 and − 1.54.	[65]
MC-LR MC-RR	• The difference between MC-RR and –LR in their hydrophobicity in acidic condition may be important to understand the toxicity difference to animals between the two toxins, not for the analytical method only	/	/	[65, 66]
MC-LR	below pH = 4: • The rate of degradation accelerated in correlation with the pKa of the carboxylic acid functionality present in MC-LR	• A slow slight MC-LR degradation	• A slow slight MC-LR degradation (moderately alkaline)	[67]
MC-LR + MC-LW	pH 3: • Concentration of 0.25g/L SPM (Suspended Particulate Matter) solution where most microcystins spiked into were adsorbed (>95%) onto solids • A linear isotherm effectively described the adsorption of MC-LR and MC-LW.	Ph 7: • Concentration of 0.25 g/L SPM solution where most microcystins spiked into were adsorbed (>85%) onto solids • a linear isotherm effectively described the adsorption of MC-LR and MC-LW.	Ph 13: • A decrease in the adsorbed microcystin proportion from 8 to 29% • A decrease in the adsorbed microcystin proportion from 38 to 47% for MC-LW • The adsorption decreased significantly with increasing pH, aligning with the pH-dependent hydrophobicity.	[68]

	Acidic	Neutral 6.5 to 7.5	Alkaline	ref
Microcystis	/	/	Ph about 8.80: • a higher cell abundance of Cyanobacteria was found	[69]
M. aeruginosa	pH 5.0: • the extracellular MC-LR concentrations in *M. aeruginosa* were low after being exposed to either naphthalene or phenanthrene,	/	pH 10.0 • the intracellular MC-LR levels were high after being exposed to either naphthalene or phenanthrene, • MC-LR production of *M. aeruginosa* was affected by the pH value	[70]
Microcystis	• Acidic pH improved the *Microcystis* unicells flocculation because they clump together to form flocs which can be easily removed by filtration or sedimentation. • Regulating the pH is an environmentally friendly and economical process to remove *Microcystis* colonies by coagulants.	/	/	[71]
M. aeruginosa *M. wesenbergii*	• if the pH was acidic to neutral, they had satisfactory removal of organic matter representative of humic acid	• Polytitanium coagulants, polytitanium chloride (PTC), and polytitanium sulfate (PTS) were used and they were highly dependent on the solution pH	• Polytitanium chloride (PTC) and polytitanium sulfate (PTS) were able to remove turbidity and algae by flocculation and flushing.	[72]

Table 3.
pH effects on Microcystis *and its toxins.*

are characterized by very low light demand and low-maximum specific growth rates in comparison to other strains studied. And their growth was similar under different light conditions.

Controlling the excessive growth of *M. aeruginosa* is a matter of great concern, many algaecides were used in correlation with natural light to inhibit and control *Microcystis* proliferation, such as $AgBiO_3$ can inhibit the proliferation of *Microcystis* due to its high photocatalytic activity, which irreversibly damages the cell walls and membranes [83] and Zn–Fe LDHs that also had a strong degradation and inhibition effect under visible light and achieved a removal rate of 80% by damaging the cell membrane and degraded cell inclusions due to its visible light catalytic activity [84]. In addition, the nanocomposite Ag/AgCl@g-C3N4@UIO-66 (NH2) can efficiently inactivate cyanobacteria under visible light [85].

	Light intensity	Light Effect	Ref.
M. aeruginosa	Summer season	• *M. aeruginosa* growth rates had a significant correlation with the extinction coefficient and mean daily photosynthetically active radiation irradiance. • The amount of light supplied to the water column is the most important factor controlling *M. aeruginosa* growth.	[74–77]
Microcystis	>10 µEinsteins m^{-2} sec^{-1}	• *Microcystis* growth was generally inhibited.	[77]
M. aeruginosa	• 0.52 to 1.32 day^{-1} • (UV-008)	• Maximum specific growth rate was the most sensitive	[77]
Aphanocapsa incerta (formerly *Microcystis incerta*)	• (UV-003)	• Was less sensitive	[77]
M. aeruginosa	• 10 to 100 µmol m^{-2} sec^{-1}	• A high tolerance for extreme light conditions and its favorable range for growth-	[78]
M. aeruginosa	• under 100 µmol m^{-2} sec^{-1} • under 500 and 1000 µmol m^{-2} sec^{-1}	• Cell growth was promoted. • An increase in the reactive oxygen species (ROS) levels • A reduced photosynthetic pigment content and photosystem II (PSII) • A gradual increase in 1O2 levels • So a growth inhibition.	[53]
Microcystin	• Up to 60 µmol m-^{2} sec^{-1} • greater than 100 µmol m^{-2} sec-^{1}	• The total microcystin production rate increased. • A positive effect of light on microcystin production and content of *Microcystis* strain PCC 7806 • The maximum growth rate is reached • The total microcystin production rate started to decrease • The microcystin production is inhibited.	[79, 80].
M. aeruginosa	1. Red light (620–700 nm), 2. white (400–700 nm) 3. red (620–700 nm)	1. In monocultures and cocultures, white light color favored the growth of *M. aeruginosa* more than other species 2. Light color mainly affected the absorption flux of reaction center (ABS/RC) in photosynthetic system II (PSII) and its potential photosynthetic capacity (Fv/Fm). 3. The red light was the suitable color to enhance its growth (Fv/Fm)	[81]

Table 4.
The effect of light changes on Microcystis *sp. and its toxins.*

As light intensity has been suggested to play a role in MC production, Chaffin et al. [86] have found that High light intensities (300 mmol photons/m2/s) with nitrogen

(N) enrichment resulted in greater cyanobacterial biovolumes and MC concentrations than lower light intensities (30 and 3 mmol photons/m2/s) and they suggest that high light intensities enhance MC production during elevated N concentrations. Another significant interaction exists between *M. aeruginosa* polysaccharide content and both light intensity and nitrogen concentration.

2.4 Salinity effect

Worldwide, aquatic ecosystems are experiencing higher salinity levels due to long-term drought, anthropogenic activities, climate change, changes in water flow, and rising sea levels [87, 88]. Presently, the rise in salinity represents one of the most crucial environmental changes, and it is essential to anticipate how this alteration will impact *Microcystis* and microcystin production in harmful strains. The previous study aimed to determine the effects of salinity on *M. aeruginosa* algae, a dominant algal species, which is known to release toxins that can contaminate drinking water sources [89, 90].

Microcystis blooms usually develop in freshwater environments due to their sensitivity to salt. Enhancing salinity levels in freshwater reservoirs has been proposed as an effective method to curb the growth of *M. aeruginosa* [91, 92]. However, some strains of *M. aeruginosa* can tolerate low to medium levels of salinity [93–95]. In 2021, Sun et al. [2] demonstrated that salinity restricts the growth of other phytoplankton, while the rapid increase in seawater temperature and changes in nutrient availability create favorable conditions for the swift proliferation of *Microcystis sp*. Additionally, a direct correlation was observed between the growth rate and the net production rate of MCs, as evidenced by [96].

According to the literature, salinity exposure can be conducted using two main methods: shock exposure, where organisms are suddenly exposed to high saline concentrations, and graded dose exposure in laboratory settings, simulating gradual salinity conditions. Additionally, in situ studies are conducted to investigate the effects of salinity in natural environments, providing a realistic perspective on the ecological and physiological impacts of salinity variations. When organisms face minimal selective pressure, they adapt by altering gene expression within hours to days, a process termed acclimation. However, once the threshold of physiological adaptation is surpassed, survival necessitates the development of new mutations that provide resistance [97, 98]. The mechanism of salt tolerance in *Microcystis sp.* includes the active extracellular transport of toxic inorganic ions and the accumulation of compatible solutes, known as osmoprotectants [99]. Research involving genomic inspection and molecular genetic analyses has revealed that the synthesis of sucrose, one of the well-known compatible solutes, is a key factor in conferring salt tolerance to *M. aeruginosa*, this process is critical for the organism's ability to cope with saline conditions [95, 100–103].

Several studies have documented the impacts of salinity on M. aeruginosa. Increased salinity levels resulted in decreased growth rate, reduced cell size, reduced biovolume, enhanced mucilage production, and reduced intercellular space [51, 96, 104–106]. The fluctuation in salinity, whether an increase or decrease, induced stress within the algae, causing an increase in total lipid content, which acted as an energy reserve until favorable conditions returned [107].

M. aeruginosa could tolerate salinity levels up to 4.0 g NaCl L^{-1}. However, as salinity increased, both cell density and chlorophyll content declined. The increase in metabolites might be a response to counteract osmotic stress. The rise in malondialdehyde (MDA) under salt stress suggests that excess reactive oxygen species (ROS) were damaging the membrane structure. It was inferred that M. aeruginosa under salinity stress accumulated a more efficient antioxidant enzyme system, including superoxide dismutase (SOD) and catalase (CAT). However, the activities of these antioxidant enzymes decreased sharply under hypersaline conditions, which could cause ROS to exceed critical levels and pose a serious threat to the cells [90]. The salinity tolerance of *M. aeruginosa*, focusing on growth rate, microcystin production, and cellular responses, has been studied extensively. Results indicate that *M. aeruginosa* maintains stable growth and microcystin production at salinity levels up to 9.8 g/l (Numerous strains of *M. aeruginosa* are capable of withstanding salinities ≤10 g NaCl L^{-1}) [93, 108–110], this stability reflects the high salt tolerance characteristic of freshwater Cyanobacteria. Nonetheless, when salinity exceeds 10 g/L, the limits of osmoregulation are surpassed. This leads to a reduction in cell size, an increase in extracellular microcystin concentrations, and, ultimately, growth inhibition. The extracellular microcystin concentration showed no significant change at salinity levels up to 10 g/L. At salinity levels of 12.5 and 15 g/L, an abrupt cessation of growth was observed, with extracellular microcystins surging notably and making up nearly 100% of the total microcystins [94]. Other authors proved that *M. aeruginosa* demonstrates the ability to tolerate elevated salinity levels through infrequent spontaneous mutations that arise during the growth of cultures subjected to lethal salinity concentrations. This acclimatization process enables survival at salt concentrations beyond the initial lethal threshold. Furthermore, successive selective mutations can enhance the lethal dose by up to 1.5 times the original amount [32]. Salt-shock experiments demonstrate temporary survival and continued microcystin production at higher salinities (10 gl^{-1}), highlighting adaptive mechanisms in response to environmental stressors. [94] After exposure to salt shocks at 15 and 17.5 g/L, there was a decline in growth rates and microcystin production. Despite this, *M. aeruginosa* persisted in growth and microcystin production for at least one week.

Bormans et al. [106] showed that a sudden increase in salinity (S ≥ 20) over 5–6 days had both morphological and physiological impacts on *M. aeruginosa*, resulting in larger colony sizes and decreased intercellular spacing, which enhanced buoyancy and reduced predation. The study revealed variable responses among strains in terms of cell size, mucilage production, and autofluorescence, with mucilage offering partial protection against salinity stress, suggesting potential survival in estuarine environments and implications for estuarine ecosystems and shellfish industries.

Microcystis, typically a freshwater species, though higher salinity can disrupt osmotic potential and sodium export, inhibiting growth [111, 112]. Adaptability to salt stress is crucial, with nitrogen availability enhancing salt tolerance and calcium supplementation mitigating damage from salt stress [94, 103]. The ability of *Microcystis* to survive in oligohaline to low mesohaline conditions (5–18‰) suggests that low nitrogen levels might be elevated.

2.5 Nutrient's effect

Freshwater bodies are frequently overloaded with nitrogen and phosphorus due to intensive anthropogenic activities, including domestic sewage, agricultural fertilizers, industrial discharges, and fossil fuel combustion. These nutrient inputs promote the global proliferation of harmful cyanobacterial blooms (HABs) [113, 114]. HABs are generally linked to nutrient-rich waters with limited water circulation. [8, 115–117]

Elevated phosphorus concentrations have been shown to increase microcystin production per cell in specific cyanobacteria [116, 118]. Nonetheless, nitrogen (N) also plays a crucial role in the formation of toxic blooms by non-N-fixing cyanobacteria, including *Microcystis* sp. [119].

2.5.1 Nitrogen effect

In eutrophic water bodies, nitrogen nutrients exist in multiple forms, including both inorganic and organic nitrogen, resulting from the diverse influx of substantial volumes of agricultural runoff, livestock waste, urban effluent, and industrial wastewater [21, 120–122].

Nitrogen (N), a vital macronutrient, plays a significant role in cyanobacterial growth and various physiological processes, including the synthesis of microcystins (MCs) [123]. It is also a crucial nutrient for *Microcystis* in temperate and subtropical lakes and reservoirs, significantly influencing photosynthesis and the biosynthesis of essential macromolecules such as proteins, nucleic acids, and chlorophyll, which are necessary for its growth and metabolic functions [124, 125]. Nitrogen is a key constituent of pigments involved in photosynthesis [126, 127]. In freshwater systems, nitrate and ammonium are the predominant nitrogen sources, these variabilities influence growth rates, photosynthetic health, MC concentrations, and the composition of total MCs and their congeners [21, 128–131].

Ammonium served as the principal nitrogen source for the *Microcystis* bloom in the studies by [132, 133]. Elevated levels of urea-N are now recognized as a crucial factor influencing the survival and distribution of aquatic organisms, sometimes surpassing inorganic nitrogen concentrations [125, 134–136]. Nitrate (NO^{3-}) is the dominant form of dissolved inorganic nitrogen in most lakes and reservoirs and can be taken up by *Microcystis* for development [137, 138]. Cells using NO^{3-} must induce nitrate reductase and expend extra energy, but NO^{3-} can be more readily accumulated within cells than other nitrogen forms [139]. In comparison, ammonium can be directly transported into algal cells and assimilated with minimal energy expenditure. However, NH^{4+} concentrations in natural water bodies are typically low, and excess cellular NH^{4+} is toxic to algal cells [140] due to the diffusion of ammonia into cells, causing photodamage [140] and pH imbalances [141], which ultimately suppress growth. Elevated concentrations of ammonia not only inhibited the uptake of other nutrients but also suppressed nitrogen assimilation, resulting in a significant decrease in primary productivity [127, 142]. Elevated nitrate concentrations can cause the accumulation of intracellular nitrite, generated by nitrate reductase activity, which in turn can inhibit algal growth [143, 144]. Due to its small size, the organic nitrogen molecule urea-N penetrates phytoplankton cells more readily than other organic nitrogen compounds [145]. Inside the cell, urea is broken down into ammonia and carbon dioxide, which are then incorporated into amino acids and proteins. This conversion involves the enzyme urease and generally requires additional energy, except in certain Chlorophyta species [125, 146–148]. When levels of inorganic nitrogen and urea are notably reduced or depleted, algae can also assimilate other organic nitrogen sources, primarily amino acids such as arginine (Arg), cysteine (Cys), and leucine (Leu) [149, 150], as well as those resulting from extracellular deamination, including lysine (Lys), serine (Ser), and alanine (Ala) [21, 151].

Previous studies have shown varying effects of nitrogen concentrations on Cyanobacteria blooms. Chaffin and Bridgeman [152] reported that low nitrogen concentrations promote cyanobacterial blooms. A study made in Algeria in 2007 proves that physical and chemical data measured at the Chaffia reservoir, Algeria, during sampling in October 2007, are particularly valuable, as these were recorded in conjunction with the presence of dense cell concentrations of the morphospecies *Microcystis* sp. *M. aeruginosa* is dominant under the following conditions: surface temperature of 23°C, alkaline pH of 8.6, with high orthophosphate content (180 μg L^{-1}) and low nitrate content (12 μg L^{-1}) [153]. Conversely, Cai and Tang [125] found that elevated nitrogen levels were responsible for a *Microcystis densa* bloom. Additionally, low nitrogen concentrations favor the growth of N2-fixing Cyanobacteria over non-N2-fixing species [128, 154, 155]. Under nitrogen-deficient conditions, *Microcystis* cells undergo various physiological changes that affect their sinking properties. Nitrogen deficiency in cyanobacteria leads to an excess of carbon, which accumulates in storage granules such as polyhydroxyalkanoate and glycogen, increasing cell density [156–159]. Proteomic analysis showed that proteins involved in starch and sucrose metabolism were differentially regulated, resulting in reduced glucose-1-phosphate (G1P) entry into the glycolysis pathway and decreased pyruvate production, decreased glycolysis was linked to reduced photosynthetic activities and inorganic carbon fixation [157, 160]. These findings suggest that nitrogen deficiency induced carbohydrate accumulation due to the slowed catabolic consumption of carbohydrates [161].

2.5.2 Phosphorus effect

Phosphorus is introduced into lakes from the widespread use of phosphorus-containing fertilizers and washing agents in agriculture, as well as from industrial wastewater and municipal sewage discharges; this influx of phosphorus greatly enhances the proliferation of cyanobacterial blooms [162, 163]. Sun et al. [2] and Kong et al. [164] noted that *Microcystis* has a higher maximum phosphorus uptake rate and greater phosphorus binding capacity compared to other algae. *M. aeruginosa* can use various phosphorus sources, giving it a competitive advantage within the phytoplankton community [165]. They possess several adaptive traits that enhance its competitive success, including gas vesicles [166], the ability to consume phosphorus in excess and store it as polyphosphate granules, tolerance to high light intensities, and the formation of large colonies encased in mucilage, which reduces grazing [39, 167].

Phosphorus is crucial for the formation of ribosomes, DNA, and cell membranes, synthesizing nucleic acids, phospholipids, and various biochemical intermediates facilitating algal growth in aquatic environments [168–170]. Phosphorus occurs in many forms, with dissolved inorganic phosphate (DIP) being the primary form directly utilized by microorganisms [165]. Phosphorus loading is a key determinant of bloom size and *Microcystis* biomass [171], which is why management strategies often focus on phosphorus [172]. Organophosphates (DOP) constitute a major portion of the phosphorus sources in many ecosystems and must be converted to dissolved inorganic phosphate before they can be used by plankton [165]. DOP is a more favorable phosphorus source for *M. aeruginosa* growth compared to three organophosphates: glucose-6-phosphate, β-glycerol phosphate, and glyphosate.

Phosphorus (P) levels are a crucial regulatory factor for the occurrence of *Microcystis* blooms (MCBs) and the synthesis of microcystins (MCs). The biosynthesis of MCs in cyanobacterial cells requires significant energy, which in turn demands high phosphorus availability [165, 173]. Additionally, P influences *Microcystis* resistance to external stress by adjusting cellular physiological processes [174–176]. Spiramycin inhibited *Microcystis* growth at high P levels (7.1 mg/L) but stimulated growth at low P levels (0.7 mg/L) by altering photosynthesis, transcription, and cell division [177]. *M. aeruginosa* can grow at a low DIP concentration of 0.02 mg P/L, but its growth rate slows and photosynthetic activity declines. At such a low concentration of phosphorus, energy metabolism and cell synthesis are inadequate, affecting processes like energy recovery, nucleic acid metabolism, and phospholipid synthesis for membranes and chlorophyll [165, 178]. Higher phosphate levels correlate positively with increased MC content in natural ecosystems, suggesting that more phosphorus enhances toxin production [179]. DOP contributes minimally to toxin production due to its limited uptake by *M. aeruginosa*, which also experiences inhibited growth and reduced MC levels. Elevated DOP concentrations (0.6 and 1.0 mg P/L) suppress *M. aeruginosa* growth, leading to fewer cells but higher single-cell toxin production [165, 180]. The availability of phosphate influences the physiological metabolism of algal cells, acting as a substrate for ATP phosphorylation and affecting nucleotide synthesis [181, 182]. Extracellular alkaline phosphatase (AP) serves as a biochemical indicator of phosphorus limitation. When ambient orthophosphate levels are low, AP is expressed on the cell surface, cleaving various phosphomonoesters into bioavailable phosphorus for direct uptake by phytoplankton [183]. Increased phosphorus enhances the growth and microcystin production of *M. aeruginosa*. This suggests that in bloom conditions, reducing phosphorus in eutrophic waters could decrease growth and microcystin production, thus reducing toxicity [184]. Nalewajko and Murphy [185] observed that *Microcystis* growth declined under phosphorus-limited conditions. In such conditions, the biomass of *Microcystis* was restricted, and the protein content per cell decreased in vivo [186]. *Microcystis* has evolved several adaptive strategies to handle phosphorus-limited conditions. They store inorganic phosphorus as polyphosphate (PolyP) when it is abundant and break it down during phosphorus stress, allowing them to endure extended periods of dissolved inorganic phosphorus (DIP) deficiency [187, 188]. At low phosphorus concentrations, colonial *Microcystis* strains outperformed unicellular strains, as they required less phosphorus for growth. These colonial strains also exhibit a higher affinity for low Pi levels, a high phosphorus cell quota, and superior photosynthetic efficiency, enabling them to sustain normal growth for extended periods under fluctuating phosphorus conditions. Indeed, mucilage may enhance the physiological traits of *Microcystis* species, providing colonial strains with a competitive advantage in nutrient acquisition within natural water columns [189].

3. Conclusions

The growth and toxin production of *Microcystis* sp. are intricately influenced by various environmental parameters. Key factors include nutrient concentrations, particularly nitrogen and phosphorus, which are critical for Cyanobacteria proliferation. Higher levels of both nutrients often lead to eutrophication, creating favorable conditions for *Microcystis* sp. blooms. Additionally, water temperature plays a crucial role as warmer temperatures enhance the metabolic rates of these Cyanobacteria, promoting rapid growth and increased toxin production. The stability of the water column, influenced by factors such as wind and stratification, affects nutrient availability and the physical mixing of water layers, which can either support or limit algal bloom formation. Furthermore, light intensity and duration impact photosynthesis and growth rates, with optimal conditions fostering higher biomass. The interplay between these environmental parameters can lead to synergistic effects, exacerbating the frequency and intensity of *Microcystis* blooms and associated microcystin production. Understanding these interactions is essential for managing water quality and mitigating the risks posed by harmful algal blooms.

Author details

Hichem Nasri*, Loubna Ghodbane, Amina Amrani, Chaima Aounallah, Fateh Sedrati, Maaref Hanen, Zineb Labidi, Zakia Hammoum, Nassima Beldjoudi and Hadjer Zaidi
Faculty of Life and Nature Sciences, Laboratory of Biodiversity and Ecosystems Pollution, University of Chadli Bendjedid El-Taref, Algeria

*Address all correspondence to: h.nasri@atrss.dz

References

[1] Danielsdottir MG, Brett MT. Arhonditsis GB phytoplankton food quality control of planktonic food web processes. Hydrobiologia. 2007;**589**:29-41

[2] Sun B, Tang C, Yang N, He P. Composition and variation of phytoplankton communities during *Microcystis* bloom in an artificial lagoon of Hangzhou Bay, China. Aquatic Ecology. 2021;**55**:467-481. DOI: 10.1007/s10452-021-09838-5

[3] Lu T, Zhang Q, Zhang Z, Hu B, Chen J, Chen J, et al. Pollutant toxicology with respect to microalgae and cyanobacteria. Journal of Environmental Sciences. 2021;**99**:175-186

[4] Igwaran A, Kayode AJ, Moloantoa KM, Khetsha ZP, Unuofin JO. Cyanobacteria harmful algae blooms: Causes, impacts, and risk management. Water, Air, & Soil Pollution. 2024;**235**(1):71. DOI: 10.1007/s11270-023-06782-y

[5] Hopes A, Mock T. Evolution of microalgae and their adaptations in different marine ecosystems. eLS. 2015;**3**:1-9

[6] Lauritano C, Rizzo C, Giudice AL, Saggiomo M. Physiological and molecular responses to main environmental stressors of microalgae and bacteria in polar marine environments. Microorganisms. 2020;**8**(12):1957

[7] Kholssi R, Stefanova S, González-Ortegón E, Araújo CV, Moreno-Garrido I. Population and functional changes in a multispecies co-culture of marine microalgae and cyanobacteria under a combination of different salinity and temperature levels. Marine Environmental Research. 2024;**193**:106279. DOI: 10.1016/j.marenvres.2023.106279

[8] Oliver RL, Ganf GG. In: Whitton BA, Potts M, editors. Freshwater Blooms. The Ecology of Cyanobacteria, their Diversity in Time and Space. Dordrecht, The Netherlands: Kluwer Academic; 2000. pp. 149-194

[9] Fogg G, Stewart WDP, Fay P, Walsby AE. The Blue-Green Algae. London: Academic Press; 1973

[10] Metcalf JS, Banack SA, Wessel RA, Lester M, Pim JG, Cassani JR, et al. Toxin analysis of freshwater cyanobacterial and marine harmful algal blooms on the west coast of Florida and implications for estuarine environments. Neurotoxicity Research. 2021;**39**:27-35. DOI: 10.1007/s12640-020-00248-3

[11] Massey IY, Al Osman M, Yang F. An overview on cyanobacterial blooms and toxins production: Their occurrence and influencing factors. Toxin Reviews. 2020;**41**(1):326-346

[12] Al Shehhi MR, Gherboudj I, Ghedira H. An overview of historical harmful algae blooms outbreaks in the Arabian seas. Marine Pollution Bulletin. 2014;**86**(1-2):314-324

[13] Pal M, Yesankar PJ, Dwivedi A, Qureshi A. Biotic control of harmful algal blooms (HABs): A brief review. Journal of Environmental Management. 2020;**268**:110687

[14] Ng PH, Cheng TH, Man KY, Huang L, Cheng KP, Lim KZ, et al. Hydrogen peroxide as a mitigation against Microcystis sp. Bloom. Aquaculture. 2023;**577**:739932. DOI: 10.1016/j.aquaculture.2023.739932

[15] Ho JC, Michalak AM, Pahlevan N. Widespread global increase in intense lake phytoplankton blooms since the 1980s. Nature. 2019;**574**:667-670. DOI: 10.1038/s41586-019-1648-7

[16] Zhang J, Wang N, Zhang Z, Gao Y, Dong J, Gao X, et al. Combined effects of toxic *Microcystis aeruginosa* and high pH on antioxidant responses, immune responses, and apoptosis of the edible freshwater bivalve *Corbicula fluminea*. Ecotoxicology and Environmental Safety. 2024;**280**:116568. DOI: 10.1016/j.ecoenv.2024.116568

[17] Kim W, Park Y, Kim M, Cha Y, Jung J, Jeon CO, et al. Sustainable control of *Microcystis aeruginosa,* a harmful cyanobacterium, using *Selaginella tamariscina* extracts. Ecotoxicology and Environmental Safety. 2024;**277**:116-375. DOI: 10.1016/j.ecoenv.2024.116375

[18] Wan X, Steinman AD, Gu Y, Zhu G, Shu X, Xue Q, et al. Occurrence and risk assessment of microcystin and its relationship with environmental factors in lakes of the eastern plain ecoregion, China. Environmental Science and Pollution Research. 2020;**27**:45095-45107. DOI: 10.1007/s11356-020-10384-0

[19] Yang Z, Kong F, Shi X, Zhang M, Xing P, Cao H. Changes in the morphology and polysaccharide content of *Microcystis aeruginosa* (cyanobacteria) during flagellate grazing1. Journal of Phycology. 2008;**44**:716-720

[20] Zhang X, Chen C, Ding J, Hou A, Li Y, Niu Z, et al. The 2007 water crisis in Wuxi, China: Analysis of the origin. Journal of Hazardous Materials. 2010;**182**:130-135

[21] Zuo Z, Yang L, Chen S, Ye C, Han Y, Wang S, et al. Effects of nitrogen nutrients on the volatile organic compound emissions from *Microcystis aeruginosa*. Ecotoxicology and Environmental Safety. 2018;**161**:214-220. DOI: 10.1016/j.ecoenv.2018.05.095

[22] Zhang Q, Song Q, Wang C, Zhou C, Lu C, Zhao M. Effects of glufosinate on the growth of and microcystin production by *Microcystis aeruginosa* at environmentally relevant concentrations. Science of the Total Environment. 2017;**575**:513-518

[23] Crettaz-Minaglia M, Fallico M, Aranda O, Juarez I, Pezzoni M, Costa C, et al. Effect of temperature on microcystin-LR removal and lysis activity on Microcystis aeruginosa (cyanobacteria) by an indigenous bacterium belonging to the genus Achromobacter. Environmental Science and Pollution Research. 2020;**27**:44427-44439. DOI: 10.1007/s11356-020-09901-y

[24] Medrano EA, Uittenbogaard RE, Pires LD, Van De Wiel BJH, Clercx HJH. Coupling hydrodynamics and buoyancy regulation in *Microcystis aeruginosa* for its vertical distribution in lakes. Ecological Modelling. 2013;**248**:41-56. DOI: 10.1016/j.ecolmodel.2012.08.029

[25] Cyr H, Curtis JM. Zooplankton community size structure and taxonomic composition affects size-selective grazing in natural communities. Oecologia. 1999;**118**(3):306-315

[26] Kearns KD, Hunter MD. Green algal extracellular products regulate antialgal toxin production in a cyanobacterium. Environmental Microbiology. 2001;**2**(3):291-297

[27] Wu X, Kong F. Effects of light and wind speed on the vertical distribution of *Microcystis aeruginosa* colonies of different sizes during a summer bloom. International Review of Hydrobiology. 2009;**94**(3):258-266

[28] Yamamoto Y, Shiah FK, Chen YL. Importance of large colony formation in bloom-forming cyanobacteria to dominate in eutrophic ponds. Annales de Limnologie – International Journal of Limnology. 2011;**47**:167-173

[29] Ma J, Brookes JD, Qin B, Paerl HW, Gao G, Wu P, et al. Environmental factors controlling colony formation in blooms of the cyanobacteria *Microcystis* spp. in Lake Taihu, China. Harmful Algae. 2014;**31**:136-142. DOI: 10.1016/j.hal.2013.10.016

[30] Bouaïcha N, Miles CO, Beach DG, Labidi Z, Djabri A, Benayache NY, et al. Structural diversity, characterization and toxicology of microcystins. Toxins. 2019;**11**:714. DOI: 10.3390/toxins11120714

[31] Paerl HW, Huisman J. Blooms like it hot. Science. 2008;**320**:57-58

[32] Li L, Chen X, Huang Y, Shen Y, Liu S, Lu J, et al. The salt tolerance of the freshwater cyanobacterium *Microcystis* depends on nitrogen availability. Science of the Total Environment. 2021;**777**:146-186. DOI: 10.1016/j.scitotenv.2021.146186

[33] Li H, Ai H, Kang L, Sun X, He Q. Simultaneous *Microcystis* algicidal and microcystin degrading capability by a single acinetobacter bacterial strain. Environmental Science & Technology. 2016a;**50**:11903-11911

[34] Li M, Zhu W, Gao L, Lu L. Changes in extracellular polysaccharide content and morphology of *Microcystis aeruginosa* at different specific growth rates. Journal of Applied Physics. 2013;**25**:1023-1030

[35] Lürling M, Roessink I. On the way to cyanobacterial blooms: Impact of the herbicide metribuzin on the competition between a green alga (*Scenedesmus*) and a cyanobacterium (*Microcystis*). Chemosphere. 2006;**65**:618-626

[36] Mello MME, Soares MCS, Roland F, Lürling M. Growth inhibition and colony formation in the cyanobacterium *Microcystis aeruginosa* induced by the cyanobacterium *Cylindrospermopsis raciborskii*. Journal of Plankton Research. 2012;**34**:987-994

[37] Wu Q, Zhang X, Jia S, Li J, Li P. Effects of the cultivable bacteria attached to *Microcystis* colonies on the colony size and growth of *Microcystis*. Journal of Freshwater Ecology. 2019;**34**:663-673

[38] Xiao M, Li M, Reynolds CS. Colony formation in the cyanobacterium *Microcystis*. Biological Reviews. 2018;**93**:1399-1420

[39] Yang Z, Kong F, Shi X, Zhang M, Xing P, Cao H. Changes in the morphology and polysaccharide content of *Microcystis aeruginosa* (cyanobacteria) during flagellate grazing. Journal of Phycology. 2008;**44**(3):716-720. DOI: 10.1111/j.1529-8817.2008.00502.x

[40] Ma X, Li M, Jiang E, Pan B, Gao L. Humic acid inhibits colony formation of the cyanobacterium *Microcystis* at high level of iron. Chemosphere. 2021;**281**:130742

[41] Tarafdar L, Mohapatra M, Muduli PR, Kumar A, Mishra DR, Rastogi G. Co-occurrence patterns and environmental factors associated with rapid onset of *Microcystis aeruginosa* bloom in a tropical coastal lagoon. Journal of Environmental Management. 2023;**325**:116580

[42] Salmaso N. Effects of climatic fluctuations and vertical mixing on the interannual trophic variability of Lake Garda, Italy. Limnology and Oceanography. 2005;**50**(2):553-565. DOI: 10.4319/lo.2005.50.2.0553

[43] Minasyan A, Christophoridis C, Wilson AE, Zervou SK, Kaloudis T, Hiskia A. Diversity of cyanobacteria and the presence of cyanotoxins in the epilimnion of Lake Yerevan (Armenia). Toxicon. 2018;**150**:28-38. DOI: 10.1016/j.toxicon.2018.04.021

[44] Eloff JN. Autecological studies on Microcystis. In: Carmichael WW, editor. The Water Environment, Algal Toxins and Health. New York: Plenum Press; 1980. pp. 71-96

[45] Watanabe MF, Oishi S. Effects of environmental factors on toxicity of a cyanobacterium (*Microcystis aeruginosa*) under culture conditions. Applied and Environmental Microbiology. 1985;**49**(5):1342-1344. DOI: 0099-2240/85/051342-03$02.00/0

[46] Albay M, Matthiensen A, Codd GA. Occurrence of toxic blue-green algae in the Kucukcekmece lagoon (Istanbul, Turkey). Environmental Toxicology: An International Journal. 2005;**20**(3):277-284. DOI: 10.1002/tox.20118

[47] Lei L, Dai J, Lin Q, Peng L. Competitive dominance of *Microcystis aeruginosa* against *Raphidiopsis raciborskii* is strain- and temperature-dependent. Knowledge and Management of Aquatic Ecosystems. 2020;**421**:36. DOI: 10.1051/kmae/2020023

[48] O'Neil JM, Davis TW, Burford MA, Gobler CJ. The rise of harmful cyanobacteria blooms: The potential roles of eutrophication and climate change. Harmful Algae. 2012;**14**:313-334. DOI: 10.1016/j.hal.2011.10.027

[49] Xiao M, Willis A, Burford MA. Differences in cyanobacterial strain responses to light and temperature reflect species plasticity. Harmful Algae. 2017;**62**:84-93. DOI: 10.1016/j.hal.2016.12.008

[50] Robarts RD, Zohary T. Temperature effects on photosynthetic capacity, respiration, and growth rates of bloom-forming cyanobacteria. New Zealand Journal of Marine and Freshwater Research. 1987;**21**:391-399. DOI: 10.1080/00288330.1987.9516235

[51] Sampognaro L, Eirín K, Martínez de la Escalera G, Segura CPA, C. Kruk: Experimental evidence on the effects of temperature and salinity in morphological traits of the *Microcystis aeruginosa* complex. Journal of Microbiological Methods. 2020;**175**:105971. DOI:10.1016/j.mimet.2020.105971

[52] Yagi O, Ohkubo N, Tomioka N, Okada M. Effect of irradiance and temperature on photosynthetic activity of the cyanobacterium *Microcystis* spp. Environmental Technology. 1994;**15**(4):389-394

[53] Yang Z, Zhang M, Yu Y, Shi X. Temperature triggers the annual cycle of *Microcystis*, comparable results from the laboratory and a large shallow lake. Chemosphere. 2020;**260**:127543. DOI: 10.1016/j.chemosphere.2020.127543

[54] Ye J, Guan Y, Wu L, Wang C, Chen J, Zhou S, et al. Effects of glyphosate on microcystin-LR production and release from *Microcystis aeruginosa* at different temperatures. Environmental Science and Pollution Research. 2020;**27**:41961-41969. DOI: 10.1007/s11356-020-10185-5

[55] Duong TT, Le TPQ, Dao TS, Pflugmacher S, Rochelle-Newall E, Hoang TK, et al. Seasonal variation of cyanobacteria and microcystins in the Nui Coc reservoir, northern Vietnam. Journal of Applied Phycology. 2013;**25**:1065-1075. DOI: 10.1007/s10811-012-9919-9

[56] Merder J, Harris T, Zhao G, Stasinopoulos DM, Rigby RA, Michalak AM. Geographic redistribution of microcystin hotspots in response to climate warming. Nature Water. 2023;**1**(10):844-854. DOI: 10.1038/s44221-023-00138-w

[57] Preece EP, Hartman R. Exploring factors that affect *Microcystis* abundance in the Sacramento San Joaquin delta. Harmful Algae. 2024;**138**:102682. DOI: 10.1016/j.hal.2024.102682

[58] Tao Y, Kong FX, Cao HS, Zhang XF. Simulative recruitment of *Microcystis* from the surface sediment in Taihu lake. Journal of Lake Science. 2005;**17**:231-236

[59] Mowe MAD, Porojan C, Abbas F, Mitrovic SM, Lim RP, Furey A, et al. Rising temperatures may increase growth rates and microcystin production in tropical *Microcystis* species. Harmful Algae. 2015;**50**:88-98. DOI: 10.1016/j.hal.2015.10.011

[60] Oliver RL, Walsby AE. Direct evidence for the role of light-mediated gas vesicle collapse in the buoyancy regulation of *Anabaena flos-aquae* (cyanobacteria) 1. Limnology and Oceanography. 1984;**29**(4):879-886

[61] Zheng T, Zhou M, Yang L, Wang Y, Wang Y, Meng Y, et al. Effects of high light and temperature on *Microcystis aeruginosa* cell growth and β-cyclocitral emission. Ecotoxicology and Environmental Safety. 2020;**192**:110313. DOI: 10.1016/j.ecoenv.2020.110313

[62] Harada KI, Tsuji K, Watanabe MF, Kondo F. Stability of microcystins from cyanobacteria—III. Effect of pH and temperature. Phycologia. 1996;**35**(suppl. 6):83-88

[63] Song L, Sano T, Li R, Watanabe MM, Liu Y, Kaya K. Microcystin production of *Microcystis viridis* (cyanobacteria) under different culture conditions. Phycological Research. 1998;**46**:19-23

[64] De Maagd PGJ, Hendriks AJ, Seinen W, Sijm DT. pH-dependent hydrophobicity of the cyanobacteria toxin microcystin-LR. Water Research. 1999;**33**(3):677-680

[65] Liang G, Xie P, Chen J, Yu T. Comparative studies on the pH dependence of DOW of microcystin-RR and-LR using LC-MS. The Scientific World Journal. 2011;**11**(1):20-26

[66] McCord J, Lang JR, Hill D, Strynar M, Chernoff N. pH dependent octanol–water partitioning coefficients of microcystin congeners. Journal of Water and Health. 2018;**16**(3):340-345. DOI: 10.2166/wh.2018.257

[67] Song W, De La Cruz AA, Rein K, O'Shea KE. Ultrasonically induced degradation of microcystin-LR and-RR: Identification of products, effect of pH, formation and destruction of peroxides. Environmental Science & Technology. 2006;**40**(12):3941-3946

[68] Liu G, Qian Y, Dai S, Feng N. Adsorption of microcystin LR and LW on suspended particulate matter (SPM) at different pH. Water, Air, and Soil Pollution. 2008;**192**:67-76. DOI: 10.1007/s11270-008-9635-x

[69] AffanMA,TouliabahHES,Al-HarbiSM, Abdulwassi NI, Turki AJ, Haque MM, et al. Influence of environmental parameters on toxic cyanobacterial bloom occurrence in a Lake of Bangladesh. Rendiconti Lincei. 2016;**27**:473-481. DOI: 10.1007/s12210-016-0502-1

[70] Huang Y, Pan H, Liu H, Xi Y, Ren D. Characteristics of growth and microcystin production of *Microcystis aeruginosa* exposed to low concentrations

of naphthalene and phenanthrene under different pH values. Toxicon. 2019;**169**:103-108. DOI: 10.1016/j.toxicon.2019.09.004

[71] Tan X, Duan Z, Duan P, Parajuli K, Newman J, Shu X, et al. Flocculation of *Microcystis* unicells induced by pH regulation: Mechanism and potential application. Chemosphere. 2021;**263**:127708. DOI: 10.1016/j.chemosphere.2020.127708

[72] Tian C, Zhao YX. Dosage and pH dependence of coagulation with polytitanium salts for the treatment of Microcystis aeruginosa-laden and *Microcystis wesenbergii*-laden surface water: The influence of basicity. Journal of Water Process Engineering. 2021;**39**:101726. DOI: 10.1016/j.jwpe.2020.101726

[73] Banares-Espana E, López-Rodas V, Salgado C, Costas E, Flores-Moya A. Inter-strain variability in the photosynthetic use of inorganic carbon, exemplified by the pH compensation point, in the cyanobacterium *Microcystis aeruginosa*. Aquatic Botany. 2006;**85**(2):159-162. DOI: 10.1016/j.aquabot.2006.03.009

[74] Pavlova V, Furnadzhieva S, Rose J, Andreeva R. Effect of temperature and light intensity on the growth, chlorophyll-a, concentration and microcystin production by *Microcystis aeruginosa*. General and Applied Plant Physiology. 2010;**36**(3-4):148-158. Available from: http://www.bio21.bas.bg/ipp/

[75] Tomioka N, Imai A, Komatsu K. Effect of light availability on *Microcystis aeruginosa* blooms in shallow hypereutrophic Lake Kasumigaura. Journal of Plankton Research. 2011;**33**(8):1263-1273

[76] Yang Z, Geng L, Wang W, Zhang J. Combined effects of temperature, light intensity, and nitrogen concentration on the growth and polysaccharide content of *Microcystis aeruginosa* in batch culture. Biochemical Systematics and Ecology. 2012;**41**:130-135. DOI: 10.1016/j.bse.2011.12.015

[77] Krüger GHJ, Eloff JN. The influence of light intensity on the growth of different *Microcystis* isolates. Journal of the Limnological Society of southern Africa. 1977;**3**(1):21-25

[78] Muhetaer G, Asaeda T, Jayasanka SM, Baniya MB, Abeynayaka HD, Rashid MH, et al. Effects of light intensity and exposure period on the growth and stress responses of two cyanobacteria species: *Pseudanabaena galeata* and *Microcystis aeruginosa*. Water. 2020;**12**(2):407. DOI: 10.3390/w12020407

[79] Wiedner C, Visser PM, Fastner J, Metcalf JS, Codd GA, Mur LR. Effects of light on the microcystin content of *Microcystis* strain PCC 7806. Applied and Environmental Microbiology. 2003;**69**(3):1475-1481. DOI: 10.1128/AEM.69.3.1475-1481.2003

[80] Tonk L, Visser PM, Christiansen G, Dittmann E, Snelder EO, Wiedner C, et al. The microcystin composition of the cyanobacterium *Planktothrix agardhii* changes toward a more toxic variant with increasing light intensity. Applied and Environmental Microbiology. 2005;**71**(9):5177-5181. DOI: 10.1128/AEM.71.9.5177-5181.2005

[81] Tan X, Zhang D, Duan Z, Parajuli K, Hu J. Effects of light color on interspecific competition between *Microcystis aeruginosa* and *Chlorella pyrenoidosa* in batch experiment. Environmental Science and Pollution Research. 2020;**27**:344-352

[82] Hesse K, Dittmann E, Börner T. Consequences of impaired microcystin production for light-dependent growth and pigmentation of *Microcystis aeruginosa* PCC 7806. FEMS Microbiology Ecology. 2001;**37**(1):39-43

[83] Yu X, Zhou J, Wang Z, Cai W. Preparation of visible light-responsive AgBiO3 bactericide and its control effect on the *Microcystis aeruginosa*. Journal of Photochemistry and Photobiology B: Biology. 2010;**101**(3):265-270. DOI: 10.1016/j.jphotobiol.2010.07.011

[84] Gu N, Gao J, Wang K, Li B, Dong W, Ma Y. *Microcystis aeruginosa* inhibition by Zn–Fe–LDHs as photocatalyst under visible light. Journal of the Taiwan Institute of Chemical Engineers. 2016;**64**:189-195. DOI: 10.1016/j.jtice.2016.04.016

[85] Fan G, Zhan J, Luo J, Lin J, Qu F, Du B, et al. Fabrication of heterostructured Ag/AgCl@ g-C3N4@ UIO-66 (NH2) nanocomposite for efficient photocatalytic inactivation of *Microcystis aeruginosa* under visible light. Journal of Hazardous Materials. 2021;**404**:124062. DOI: 10.1016/j.jhazmat.2020.124062

[86] Chaffin JD, Davis TW, Smith DJ, Baer MM, Dick GJ. Interactions between nitrogen form, loading rate, and light intensity on *Microcystis* and *Planktothrix* growth and microcystin production. Harmful Algae. 2018;**73**:84-97. DOI: 10.1016/j.hal.2018.02.001

[87] White E, Kaplan D. Restore or retreat? Saltwater intrusion and water management in coastal wetlands. Ecosystem Health and Sustainability. 2017;**3**(1):e01258

[88] Ross C, Warhurst BC, Brown A, Huff C, Ochrietor JD. Mesohaline conditions represent the threshold for oxidative stress, cell death and toxin release in the cyanobacterium *Microcystis aeruginosa*. Aquatic Toxicology. 2019;**206**:203-211

[89] Pan G, Zhang MM, Chen H, Zou H, Yan H. Removal of cyanobacterial blooms in Taihu Lake using local soils. I. Equilibrium and kinetic screening on the flocculation of *Microcystis aeruginosa* using commercially available clays and minerals. Environmental Pollution. 2006;**141**(2):195-200

[90] Zhang Y, Xu Q, Xi B. Effect of NaCl salinity on the growth, metabolites, and antioxidant system of *Microcystis aeruginosa*. Journal of Freshwater Ecology. 2013;**28**(4):477-487. DOI: 10.1080/02705060.2013.782579

[91] Verspagen JM, Passarge J, Jöhnk KD, et al. Water management strategies against toxic *Microcystis* blooms in the Dutch delta. Ecological Applications. 2006;**16**:313-327

[92] Melero-Jiménez IJ, Martín-Clemente E, García-Sánchez MJ, Flores-Moya A, Bañares-España E. Adaptation of the toxic freshwater cyanobacterium *Microcystis aeruginosa* to salinity is achieved by the selection of spontaneous mutants. Phycological Research. 2019;**67**(3):192-201

[93] Otsuka S, Suda S, Li R, et al. Characterization of morphospecies and strains of the genus *Microcystis* (cyanobacteria) for a reconsideration of species classification. Phycological Research. 1999;**47**:189-197

[94] Tonk L, Bosch K, Visser PM, Huisman J. Salt tolerance of the harmful cyanobacterium *Microcystis aeruginosa*. Aquatic Microbial Ecology. 2007;**46**:117-123

[95] Tanabe Y, Yamaguchi H, Sano T, Kawachi M. A novel salt-tolerant genotype illuminates the sucrose gene evolution in freshwater bloom-forming cyanobacterium *Microcystis aeruginosa*. FEMS Microbiology Letters. 2019;**366**(15):fnz190. DOI: 10.1093/femsle/fnz190/5561441

[96] Georges des Aulnois M, Roux P, Caruana A, Réveillon D, Briand E, Hervé F, et al. Physiological and metabolic responses of freshwater and brackish-water strains of *Microcystis aeruginosa* acclimated to a salinity gradient: Insight into salt tolerance. Applied and Environmental Microbiology. 2019;**85**(21):e01614-e01619. DOI: 10.1128/AEM.01614-19

[97] Sniegowski PDA, Lenski RE. Mutation and adaptation: The directed mutation controversy in evolutionary perspective. Annual Review of Ecology, Evolution, and Systematics. 1995;**26**:553-578

[98] Hughes AL. Adaptive Evolution of Genes and Genomes. New York, NY: Oxford University Press; 1999

[99] Hagemann M. Molecular biology of cyanobacterial salt acclimation. FEMS Microbiology Reviews. 2011;**35**:87-123

[100] Kolman MA, Torres LL, Martin ML, et al. Sucrose synthase in unicellular cyanobacteria and its relationship with salt and hypoxic stress. Planta. 2012;**235**:955-964

[101] Kolman MA, Salerno GL. Sucrose in bloom-forming cyanobacteria: Loss and gain of genes involved in its biosynthesis. Environmental Microbiology. 2016;**18**:439-449

[102] Sandrini G, Huisman J, Matthijs HC. Potassium sensitivity differs among strains of the harmful cyanobacterium *Microcystis* and correlates with the presence of salt tolerance genes. FEMS Microbiology Letters. 2015;**362**:fnv121

[103] Tanabe Y, Yamaguchi H. Evolutionary history of phycoerythrin pigmentation in the water bloom-forming cyanobacterium *Microcystis aeruginosa*. bioRxiv. 2018;**4**:85508. DOI: 10.1101/485508

[104] Kruk C, Segura AM, Nogueira L, Alcántara I, Calliari D, Martínez de la Escalera G, et al. A multilevel trait-based approach to the ecological performance of *Microcystis aeruginosa* complex from headwaters to the ocean. Harmful Algae. 2017;**70**:23-36

[105] Reignier O, Bormans M, Marchand L, Sinquin C, Amzi Z, Zykwinska A, et al. Production and composition of extracellular polymeric substances (EPS) by a unicellular strain and natural colonies of *Microcystis*: Impact of salinity and nutrient stress. 2023;**15**(6):783-796. DOI: 10.21203/rs.3.rs-2818009/v1

[106] Bormans M, Legrand B, Waisbord N, Briand E. Morphological and physiological impacts of salinity on colonial strains of the cyanobacteria *Microcystis aeruginosa*. Microbiology Open. 2023;**12**(3):e1367. DOI: 10.1002/mbo3.1367

[107] Asulabh KS, Supriya G, Ramachandra TV. Effect of salinity concentrations on growth rate and lipid concentration in *Microcystis sp.*, *Chlorococcum sp*. and *Chaetoceros* sp. In: National Conference on Conservation and Management of Wetland Ecosystems. School of Environmental Sciences. Kottayam, Kerala: Mahatma Gandhi University; 2012

[108] Sellner KG, Lacouture RV, Parrish CR. Effects of increasing salinity

on a cyanobacteria bloom in the Potomac estuary. Journal of Plankton Research. 1988;**10**:49-61

[109] Atkins R, Rose T, Brown RS, Robb M. The *Microcystis* cyanobacterial bloom in the Swan River: February. Water Science and Technology. 2001;**43**:107-114

[110] Orr PT, Jones GJ, Douglas GB. Response of cultured *Microcystis aeruginosa* from the Swan River, Australia, to elevated salt concentration and consequences for bloom and toxin management in estuaries. Marine and Freshwater Research. 2004;**55**(3):277-283

[111] Taylor R, Fletcher RL, Raven JA. Preliminary studies on the growth of selected "green tide" algae in laboratory culture: Effects of irradiance, temperature, salinity and nutrients on growth rate. Botany. 2001;**44**:327-336. DOI: 10.1515/BOT.2001.042

[112] Verma E, Chakraborty S, Tiwari B, Singh S, Mishra AK. Alleviation of NaCl toxicity in the cyanobacterium *Synechococcus* sp. PCC 7942 by exogenous calcium supplementation. Journal of Applied Phycology. 2018;**30**:1465-1482. DOI: 10.1007/s10811-018-1410-9

[113] Huisman J, Codd GA, Paerl HW, Ibelings BW, Verspagen JMH, et al. Cyanobacterial blooms. Nature Reviews. Microbiology. 2018;**16**:471-483

[114] Yi Q, Chen Q, Hu L, Shi W. Tracking nitrogen sources, transformation, and transport at a basin scale with complex plain river networks. Environmental Science & Technology. 2017;**51**:5396-5403

[115] Carmichael WW. The toxins of cyanobacteria. Scientific American. 1994;**270**:78-86

[116] Rapala J, Sivonen K, Lyra C, Niemela SI. Variation of microcystin, cyanobacterial hepatotoxins, in *anabaena* spp. as a function of growth stimulation. Applied and Environmental Microbiology. 1997;**63**:2206-2212

[117] Paerl HW, Fulton RS, Moisander PH, Dyble J. Harmful freshwater algal blooms with an emphasis on cyanobacteria. Scientific World. 2001;**1**:76-113

[118] Utkilen H, Gjølme N. Iron-stimulated toxin production in *Microcystis aeruginosa*. Applied and Environmental Microbiology. 1995;**61**:797-800

[119] Davis TW, Berry DL, Boyer GL, Gobler CJ. The effects of temperature and nutrients on the growth and dynamics of toxic and non-toxic strains of *Microcystis* during cyanobacteria blooms. Harmful Algae. 2009;**8**(5): 715-725. DOI: 10.1016/j.hal.2009.02.004

[120] Aslan S, Karapinar IK. Batch kinetics of nitrogen and phosphorus removal from synthetic wastewater by algae. Ecological Engineering. 2006;**28**:64-70. DOI: 10.1016/j.ecoleng.2006.04.003

[121] Boelee NC, Temmink H, Janssen M, Buisman CJN, Wijffels RH. Nitrogen and phosphorus removal from municipal wastewater effluent using microalgal biofilms. Water Research. 2011;**45**:5925-5933

[122] Abdel-Raouf N, Al-Homaidan AA, Ibraheem IBM. Microalgae and wastewater treatment. Saudi Journal of Biological Sciences. 2012;**19**:257-275

[123] Horst GP, Sarnelle O, White JD, Hamilton SK, Kaul RB, et al. Nitrogen availability increases the toxin

quota of a harmful cyanobacterium, *Microcystis aeruginosa*. Water Research. 2014;**54**:188-198

[124] Gao G, Xia J, Yu J, Zeng X. Physiological response of a red tide alga (*Skeletonema costatum*) to nitrate enrichment, with special reference to inorganic carbon acquisition. Marine Environmental Research. 2018;**133**:15

[125] Cai Y, Tang CA. Kinetics model for predicting *Microcystis* growth based on the synergistic effect of nitrogen and phosphorus on the growth of *Microcystis densa* (cyanobacteria). Polish Journal of Environmental Studies. 2021;**30**(2):1067-1077. DOI: 10.15244/pjoes/124756

[126] Shapiro SL, Kollman VH, Campillo AJ. Energy transfer in photosynthesis: Pigment concentration effects and fluorescent lifetimes. FEBS Letters. 1975;**54**(3):358-362

[127] Tan X, Gu H, Ruan Y, Zhong J, Parajuli K, Hu J. Effects of nitrogen on interspecific competition between two cell-size cyanobacteria: *Microcystis aeruginosa* and *Synechococcus* sp. Harmful Algae. 2019;**1016**(89):101661. DOI: 10.1016/j.hal.2019.101661

[128] Erratt KJ, Creed IF, Trick CG. Comparative effects of ammonium, nitrate and urea on growth and photosynthetic efficiency of three bloom-forming cyanobacteria. Freshwater Biology. 2018;**63**(7):626-638

[129] Peng G, Martin R, Dearth S, Sun X, Boyer GL, Campagna S, et al. Seasonally-relevant cool temperatures interact with N chemistry to increase microcystins produced in lab cultures of *Microcystis aeruginosa* NIES-843. Environmental Science & Technology. 2018;**52**:4127-4136

[130] Chen Q, Wang M, Zhang J, Shi W, Mynett AE, Yan H, et al. Physiological effects of nitrate, ammonium, and urea on the growth and microcystins contamination of *Microcystis aeruginosa*: Implication for nitrogen mitigation. Water Research. 2019;**163**:114890

[131] Krausfeldt LE, Farmer AT, Castro HF, Boyer GL, Campagna SR, Wilhelm SW. Nitrogen flux into metabolites and microcystins changes in response to different nitrogen sources in *Microcystis aeruginosa* NIES-843. Environmental Microbiology. 2020;**22**(6):2419-2431. DOI: 10.1111/1462-2920.15032

[132] Lehman PW, Kendall C, Guerin MA, Young MB, Silva SR, Boyer GL, et al. Characterization of the *Microcystis* bloom and its nitrogen supply in San Francisco estuary using stable isotopes. Estuaries and Coasts. 2015;**38**:165-178

[133] Lehman PW, Kurobe T, Lesmeister S, Baxa D, Tung A, The SJ. Impacts of the 2014 severe drought on the *Microcystis* bloom in San Francisco estuary. Harmful Algae. 2017;**63**:94-108. DOI: 10.1016/j.hal.2017.01.011

[134] Probyn TA, Bernard S, Pitcher GC, Pienaar RN. Ecophysiological studies on *Aureococcus anophagefferens* blooms in Saldanha Bay, South Africa. Harmful Algae. 2010;**9**:123-133

[135] Huang WM, Bi YH, Hu ZY. Effects of fertilizer-urea on growth, photosynthetic activity and microcystins production of *Microcystis aeruginosa* isolated from Dianchi Lake. Bulletin of Environmental Contamination and Toxicology. 2014;**92**:514-519

[136] Siuda W, Kiersztyn B. Urea in lake ecosystem: The origin, concentration and distribution in relation to trophic state of The great Mazurian Lakes (Poland). Polish Journal of Ecology. 2014;**63**:110-123

[137] Burberg C, Ilic M, Petzoldt T, Von Elert E. Nitrate determines growth and protease inhibitor content of the cyanobacterium *Microcystis aeruginosa*. Journal of Applied Phycology. 2019;**31**(3):1697

[138] Amono Y, Takahashi K, Machida M. Competition between the cyanobacterium *Microcystis aeruginosa* and the diatom *Cyclotella sp*. under nitrogen limited condition caused by dilution in eutrophic lake. Journal of Applied Phycology. 2012;**24**:965

[139] Kim H, Jo BY, Kim HS. Effect of different concentrations and ratios of ammonium, nitrate, and phosphorus on growth of the blue-green alga (cyanobacterium) *Microcystis aeruginosa* isolated from the Nakdong River, Korea. Algae. 2017;**32**(4):275

[140] Collos Y, Harrison PJ. Acclimation and toxicity of high ammonium concentrations to unicellular algae. Marine Pollution Bulletin. 2014;**80**:8-23

[141] Britto DT, Kronzucker HJ. NH_4 þ toxicity in higher plants: A critical review. Journal of Plant Physiology. 2002;**159**:567-584

[142] Parker AE, Dugdale RC, Wilkerson FP. Elevated ammonium concentrations from wastewater discharge depress primary productivity in the Sacramento River and the northern San Francisco estuary. Marine Pollution Bulletin. 2012;**64**(3):574-586

[143] Chen W, Zhang Q, Dai S. Effects of nitrate on intracellular nitrite and growth of *Microcystis aeruginosa*. Journal of Applied Phycology. 2009;**21**:701-706

[144] Wang W, Liu Y, Geng L, Yang Z, Chen Y. The effect of nitrogen concentration and its interaction with light intensity on population dynamics and polysaccharide content of *Microcystis aeruginosa*. Journal of Freshwater Ecology. 2011;**26**(2):241-248. DOI: 10.1080/02705060.2011.555205

[145] Wu X, Yan Y, Wang P, Ni L, Gao J, et al. Effect of urea on growth and microcystins production of *Microcystis aeruginosa*. Bioresource Technology. 2015;**181**:72-77

[146] Baker KM, Gobler CJ, Collier JL. Urease gene sequences from algae and heterotrophic bacteria in axenic and nonaxenic phytoplankton cultures. Journal of Phycology. 2009;**45**:625-634

[147] Agostoni M, Erdner DL. Analysis of ammonium transporter and urease gene expression in *Aureoumbra lagunensis*. Harmful Algae. 2011;**10**:549-556

[148] Liu Y, Chen T, Song S, Li C. Effects of nitrogenous nutrition on growth and nitrogen assimilation enzymes of dinoflagellate *Akashiwo sanguinea*. Harmful Alage. 2015;**50**:99-106

[149] Yan H, Pan G, Zou H, Song LR, Zhang MM. Effects of nitrogen forms on the production of cyanobacterial toxin microcystin-LR by an isolated *Microcystis aeruginosa*. Journal of Environmental Science and Health. 2004;**39**:2003-2993

[150] Zuo Z, Rong Q, Chen K, Yang L, Chen Z, Peng K, et al. Study of amino acids as nitrogen source in *Chlamydomonas reinhardtii*. Phycological Research. 2012b;**60**:161-168

[151] Muñoz-Blanco J, Hidalgo-Martínez J, Cárdenas J. Extracellular deamination of L-amino acids by *Chlamydomonas reinhardtii* cells. Planta. 1990;**182**:194-198

[152] Chaffin JD, Bridgeman TB. Organic and inorganic nitrogen utilization by nitrogen-stressed cyanobacteria during

bloom conditions. Journal of Applied Phycology. 2013;**26**:299-309

[153] Nasri H, Bouaïcha N, Harche MK. A new morphospecies of *Microcystis* sp. forming bloom in the Cheffia dam (Algeria): Seasonal variation of microcystin concentrations in raw water and their removal in a full-scale treatment plant. Environmental Toxicology: An International Journal. 2007;**22**(4):347-356

[154] Paerl HW, Xu H, McCarthy MJ, Zhu G, Qin B, Li Y, et al. Controlling harmful cyanobacterial blooms in a hyper-eutrophic Lake (lake Taihu, China): The need for a dual nutrient (N & P) management. Water Research. 2011b;**45**:1973-1983

[155] Paerl HW, Otten TG. Harmful cyanobacterial blooms: causes, consequences, and controls. Microbial Ecology. 2013;**65**:995-1010

[156] Hasunuma T, Kikuyama F, Matsuda M, Aikawa S, Izumi Y, Kondo A. Dynamic metabolic profiling of cyanobacterial glycogen biosynthesis under conditions of nitrate depletion. Journal of Experimental Botany. 2013;**64**(10):2943-2954

[157] Hickman JW, Kotovic KM, Miller C, Warrener P, Kaiser B, Jurista T, et al. Glycogen synthesis is a required component of the nitrogen stress response in *Synechococcus elongatus* PCC 7942. Algal Research. 2013;**2**(2):98-106

[158] Yue D, Peng Y, Yin Q, Xiao L. Proteomic analysis of *Microcystis aeruginosa* in response to nitrogen and phosphorus starvation. Journal of Applied Phycology. 2015;**27**(3):1195-1204

[159] Deschoenmaeker F, Facchini R, Cabrera Pino JC, Bayon-Vicente G, Sachdeva N, Flammang P, et al. Nitrogen depletion in *Arthrospira* sp. PCC 8005, an ultrastructural point of view. Journal of Plankton Research. 2016;**196**(3):385-393

[160] Deschoenmaeker F, Facchini R, Leroy B, Badri H, Zhang CC, Wattiez R. Proteomic and cellular views of *Arthrospira* sp. PCC 8005 adaptation to nitrogen depletion. Microbiology. 2014;**160**:1224-1236

[161] Huang Y, Liu S, Shen Y, Hong J, Peng L, Li L, et al. Nitrogen availability affects the dynamics of *Microcystis* blooms by regulating the downward transport of biomass. Harmful Algae. 2020;**93**:101796. DOI: 10.1016/j.hal.2020.101796

[162] Conley DJ, Paerl HW, Howarth RW, Boesch DF, Seitzinger SP, Havens KE, et al. Controlling eutrophication: Nitrogen and phosphorus. Science. 2009;**323**(5917):1014-1015

[163] Burberg C, Petzoldt T, Von Elert E. Phosphate limitation increases content of protease inhibitors in the cyanobacterium *Microcystis aeruginosa*. Toxins (Basel). 2020;**12**(1):33

[164] Kong F, Cao H, Tan X. Development of research on recruitment of bloom-forming cyanobacteria and blooms forecast. Environmental Monitoring and Forewarning. 2010;**2**:1-4

[165] Xie E, Su Y, Deng S, Kontopyrgou M, Zhang D. Significant influence of phosphorus resources on the growth and alkaline phosphatase activities of *Microcystis aeruginosa*. Environmental Pollution. 2021;**268**:115807. DOI: 10.1016/j.envpol.2020.115807

[166] Chu Z, Jin X, Yang B, Zeng Q. Buoyancy regulation of *Microcystis flosaquae* during phosphorus-limited

and nitrogen-limited growth. Journal of Plankton Research. 2007;**29**(9):739-745

[167] Ghaffar S, Stevenson RJ, Khan Z. Effect of phosphorus stress on *Microcystis aeruginosa* growth and phosphorus uptake. PLoS One. 2017;**12**(3):e0174349. DOI: 10.1371/journal.pone.0174349

[168] Martin P, Dyhrman ST, Lomas MW, Poulton NJ, Van Mooy B.A. Accumulation and enhanced cycling of polyphosphate by Sargasso Sea plankton in response to low phosphorus. Proceedings. National Academy of Sciences. United States of America 2014;**111**(22):8089-8094

[169] Muscarella ME, Bird KC, Larsen ML, Placella SA, Lennon JT. Phosphorus resource heterogeneity in microbial food webs. Aquatic Microbial Ecology. 2014;**73**(3):259-272

[170] Michalak AM, Anderson EJ, Beletsky D, Boland S, Bosch NS, Bridgeman TB, et al. Record-setting algal bloom in Lake Erie caused by agricultural and meteorological trends consistent with expected future conditions. Proceedings of the National Academy of Sciences of the United States of America. 2013;**110**(16):6448-6452

[171] Ohio EPA. Ohio Lake Erie Phosphorus Task Force II: Final Report. Columbus, Ohio: Ohio Environmental Protection Agency; 2013. Available from: https://lakeerie.ohio.gov/Portals/0/Reports/Task_Force_Report_October_2013.pdf

[172] Seweryn P, Van LB, Kjeldgaard M, Russo CJ, Passmore LA, Hove-Jensen B, et al. Structural insights into the bacterial carbon–phosphorus lyase machinery. Nature. 2015;**525**(7567):68-72

[173] Topcu D, Brockmann U. Consistency of thresholds for eutrophication assessments, examples and recommendations. Environmental Monitoring and Assessment. 2021;**193**:677

[174] Liu Y, Chen S, Chen X, Zhang J, Gao BY. Interactions between *Microcystis aeruginosa* and coexisting amoxicillin contaminant at different phosphorus levels. Journal of Hazardous Materials. 2015;**297**:83-91

[175] Chen S, Liu Y, Zhang J, Gao BY. iTRAQ-based quantitative proteomic analysis of *Microcystis aeruginosa* exposed to spiramycin at different nutrient levels. Aquatic Toxicology. 2017;**185**:193-200

[176] Luo L, Li JM, Zhang Z, Yuan Y. Phosphorus influences the interaction between toxigenic *Microcystis* and chloramphenicol. Bulletin of Environmental Contamination and Toxicology. 2019;**102**:391-398

[177] Li J, Chen Y, Wan Q, Zhang M. Phosphorus level impacts luteolin effect on *Microcystis aeruginosa* growth and microcystin-pollution risk – novel perspective from correlation between exopolymers substances fractions and microcystin-production/release. Ecotoxicology and Environmental Safety. 2023;**255**:114794. DOI: 10.1016/j.ecoenv.2023.114794

[178] Li Q, Fu L, Wang Y, Zhou D, Rittmann BE. Excessive phosphorus caused inhibition and cell damage during heterotrophic growth of *Chlorella regularis*. Bioresource Technology. 2018;**268**:266-270

[179] Srivastava A, Ko SR, Ahn CY, Oh HM, Ravi AK, Asthana RK. Microcystin biosynthesis and mcyA expression in geographically distinct

Microcystis strains under different nitrogen, phosphorus, and boron regimes. BioMed Research International. 2016;(1):1-13

[180] Tilahun S, Kifle D, Zewde TW, Johansen JA, Demissie TB, Hansen JH. Temporal dynamics of intra-and extra-cellular microcystins concentrations in Koka reservoir (Ethiopia): Implications for public health risk. Toxicon. 2019;**168**(12):83-92

[181] Shi XL, Yang LY, Jiang LJ, Kong FX, Qin BQ, Gao G. Intracellular 590 phosphorus metabolism and growth of *Microcystis aeruginosa* in dark/light cycles under 591 various redox potential difference conditions. Hydrobiologia. 2006;**581**:167-176

[182] Ren L, Wang P, Wang C, Paerl HW, Wang H. Effects of phosphorus availability and phosphorus utilization behavior of *Microcystis aeruginosa* on its adaptation capability to ultraviolet radiation. Environmental Pollution. 2019;**256**:113441. DOI: 10.1016/j.envpol.2019.113441

[183] Wan L, Chen X, Denga Q, Yang L, Li X, Zhang J, et al. Phosphorus strategy in bloom-forming cyanobacteria (*Dolichospermum* and *Microcystis*) and its role in their succession. Harmful Algae. 2019;**84**:46-45

[184] Oh H-M, Lee SJ, Jang M-H, Yoon B-D. Microcystin production by *Microcystis aeruginosa* in a phosphorus-limited Chemostat. Applied and. Environmental Microbiology. 2000;**66**(1):176-179

[185] Nalewajko C, Murphy TP. Effects of temperature, and availability of nitrogen and phosphorus on the abundance of *anabaena* and *Microcystis* in Lake Biwa, Japan: An experimental approach. Limnology. 2001;**2**:45-48

[186] Ou MM, Wang Y, Cai WM. Physiological and biochemical changes in *Microcystis aeruginosa* Qutz. in phosphorus limitation. Journal of Integrative Plant Biology. 2005;**47**(6):692-702

[187] Garbisu C, Hall DO, Serra JL. Removal of phosphate by foam-immobilized *Phormidium laminosum* in batch and continuous-flow bioreactors. Journal of Chemical Technology & Biotechnology. 1993;**57**(2):181-189

[188] Ritchie RJ, Donelle AT, Larkum AWD. Phosphate limited cultures of the cyanobacterium *Synechococcus* are capable of very rapid, opportunistic uptake of phosphate. New Phytologist. 2001;**152**:189-201

[189] Shen H, Song L. Comparative studies on physiological responses to phosphorus in two phenotypes of bloom-forming *Microcystis*. Hydrobiologia. 2007;**592**:475-486

Section 2

Marine and Coastal Ecosystem Studies

Chapter 6

Summer Variability of the Zooplankton Community along the El Bibane Lagoon (Tunisia, Eastern Mediterranean)

Amira Rekik, Ahmad J. Al-Shemmari, Marc Pagno, Mohammad Ali, Hanan Al-Adeelah, Adel Naseeb, Ahmad Al-Khayat, Mohammad Boarki, Neila Annabi-Trabelsi, Wassim Guermazi, Habib Ayadi and Jannet Elloumi

Abstract

Studying zooplankton variability in coastal lagoons is crucial for environmental monitoring and preserving marine biodiversity. These organisms are highly valued as bio-indicators and helpful in identifying environmental phenomena such as eutrophication and assessing water quality. We studied the distribution of zooplankton in relation to environmental factors, phytoplankton, and ciliates in the coastal waters of the El Bibane lagoon during the summer of 2009 and 2010. Zooplankton assemblages were dominated by copepods, which represented 73–89% and 95–98% of total zooplankton abundance in summer 2009 and summer 2010, respectively. A total of 11 (summer 2009) and 21 (summer 2010) copepod species were identified in all stations, with an overwhelming abundance of *Oithona nana* in the summer of 2009 and *Oithona similis* in the summer of 2010. The prevalence of the two *Oithona* species is attributed to their adaptive strategies, which enables them to withstand environmental disturbances. Other zooplankton developed in association with an important proliferation of Cladoceran and Fish larvae, contributing 51% and 37% of total other zooplankton abundances in summer 2009 and summer 2010, respectively. The abundance of copepods or other zooplankton showed no significant correlation with phytoplankton and ciliates for both periods, suggesting an omnivorous/detritivorous diet hypothesis in El Bibane lagoon.

Keywords: El Bibane lagoon, zooplankton, phytoplankton, ciliate, environmental factors, summer season

1. Introduction

Plankton communities are viewed as being structured by a combination of factors such as climate change, abiotic properties, biotic factors [1], and regional effects (pollution) [2]. Zooplankton is a critical trophic link between phytoplankton, commercial fish stocks, and sea birds [3], and small zooplankton is an essential link between classical and microbial trophic levels [4]. The zooplankton species may respond differently to food concentration and environmental parameter variations [3]. In this context, it is important to understand the fluctuations in zooplankton abundance and their associations with physicochemical changes in the marine environment. These factors affect zooplankton at different scales [5], both directly [6] and indirectly, by shaping the abundance and repartition of their predators and prey [7], which will define populations' growth [8] and mortality rates [9]. Planktonic copepods are the major members of zooplankton in marine pelagic ecosystems [10]. They graze on nano-micro-phytoplankton and microzooplankton [11], and at the same time, they are preyed on by higher trophic levels, such as fishes. Planktonic ciliates play an important role in transferring the production of pico and nanoplankton to meso- and macroplankton [12]. Variations in the repartition of ciliates may significantly affect other components of the marine food web and thus may influence the community structure and species composition of lower and higher organisms [13]. This study deals with the El Bibane lagoon (Gulf of Gabès, Tunisia), a symbolic Mediterranean ecosystem classified as Ramsar Wetland since 2007. This shallow lagoon is under eutrophication stress. The Gulf of Gabès, situated in southern Tunisia, is a unique and ecologically significant area in the Mediterranean Sea. Unfortunately, this dynamic marine ecosystem is undergoing cultural eutrophication, characterized by nutrient enrichment from phosphorus and nitrogen salts [14]. Major contributors to nutrient pollution in the Gulf of Gabès include industrial discharges, urbanization, and agricultural activities [15]. A preceding study concerns the micro-phytoplankton communities [16]. Here, we focus on zooplankton, a crucial biological indicator of water quality and trophic status in lagoons, as they rapidly respond to environmental changes. With the objectives to (i) determine whether physicochemical properties such as water temperature, salinity, pH, and nutrient concentrations significantly impact the occurrence of the different zooplankton species; and (ii) analyze the dynamic and diversity of the zooplankton community in relation to phytoplankton and ciliates.

2. Materials and methods

2.1 Study site

The El Bibane lagoon, also known as Bhiret el Bibane, is a large lagoon of around 33 km in length by 10 km in width next to the Libyan limit. It is 10 km North of Ben Gardane city and 20 km West of Zarzis city. It is linked to the sea by a series of small channels, the largest of which is 800 m wide. This lagoon is the second largest in Tunisia after the Boughrara lagoon (**Figure 1**).

2.2 Field sampling

Samples for nutrients, phytoplankton, zooplankton, and ciliates were collected during one-day campaigns performed in summer (July 2009 and July 2010) at four

Figure 1.
Location of sampling stations along the El Bibane lagoon.

stations (**Figure 1**). Seawater samples for nutrients, phytoplankton, and ciliates were collected with a Van Dorn-type closing bottle deployed horizontally. Zooplankton samples were collected using a cylinder-conical net (30 cm aperture, 100 cm high, and 100 μm mesh size). The volume of water filtered was about 1 m^3. Back in the laboratory, samples for nutrient analyses (120 mL) were immediately filtered under a low vacuum (<50 mm Hg) through pre-combusted (500°C, 4 h) GF/F (~0.7 mm) glass fiber filters (25 or 47 mm diameter, Whatman) using glassware filtration systems. Nutriment samples (120 mL) were kept immediately upon collection at −20°C in the dark. Samples for phytoplankton were preserved with acid Lugol solution (at 3%; [17]), and alkaline Lugol solution was used for fixation of ciliates samples (at 5%; [18]). Zooplankton samples were preserved in a 2% buffered formaldehyde solution and were stained with rose Bengal to facilitate dissection. Samples for plankton were placed at 4°C in the dark for enumeration.

2.3 Physicochemical variables

Physicochemical parameters (temperature, salinity, and pH) were measured using a multi-parameter kit (Multi 340 i/SET) immediately after sampling. Water for nutrient (nitrite, nitrate, ammonium, orthophosphate, silicate, total nitrogen, and total phosphate) analyses was collected in plastic containers of 4.5 mL previously washed with distilled water. Samples were analyzed with a Bran and Luebbe type 3 autoanalyzer, and concentrations were determined colorimetrically using a UV-visible (6400/6405) spectrophotometer [19]. Analyses are independent. The automatic analysis system provides a fast and accurate analysis of these nutrients. Although each nutrient is determined differently, the method remains similar. It used colorimetry to determine the dosage of each nutrient.

2.4 Plankton enumeration

Sub-samples (50 mL) for phytoplankton and ciliates counting were analyzed under an inverted microscope using the Utermöhl method [20] after 24 h settling. Phytoplankton and ciliates species counts were carried out on the entire sedimentation chamber with 40X magnification. Phytoplankton and ciliates were identified using morphological criteria. Phytoplankton species were identified according to various keys [21, 22]. Ciliates were identified at the genus or species level after the works [23–25]. Zooplankton enumeration was performed under a vertically mounted deep focus dissecting microscope (Olympus TL 2) after being colored with Bengalrose to identify internal tissues of the different zooplankton species and also to facilitate copepod dissection such as various appendixes and leg 5 of the different species. Zooplankton species identification was identified using various keys [26–28]. Their relative frequency determined the importance value for the different species.

2.5 Statistical analyses

The environmental parameters assessed at four stations and two periods were submitted to a normalized principal component analysis (PCA) [29]. Physical-chemical variables, such as temperature, salinity, pH, nutrient concentrations, and biological parameters, were assessed by examining the projection of the plots of the extracted factors on a factorial plan consisting of the statistically significant axis of the PCA. A simple log (x + 1) transformation was applied to the data in order to stabilize variance correctly [30]. The spatiotemporal patterns of zooplankton taxonomic groups were assessed with a non-metric multidimensional scaling (NMDS) after square-root transformation of data using PRIMER v7 software. Means and standard deviations (SD) were reported when appropriate. The spatial variability of biological communities and their relationships with environmental factors were assessed using Spearman rank correlation.

3. Results

3.1 Physical, chemical, and trophic parameters

The mean and range values of the four studied stations for physical and chemical variables recorded in July 2009 and July 2010 are given in **Table 1**. Surface water temperature was slightly warmer in summer 2010 (28.68 ± 0.54°C) than in summer 2009 (28.20 ± 0.39°C) (**Table 1**). The highest temperature (29.15°C) was recorded in the summer of 2010 and the lowest one (27.74°C) in the summer of 2009 at station 3. The average salinity was 45.46 ± 0.77 psu in the summer of 2010, and 45.77 ± 0.96 psu in the summer of 2009 (**Table 1**). pH was higher in the summer of 2009 (8.35 ± 0.19) than in the summer of 2010 (8.29 ± 0.16) (**Table 1**).

The concentration of total nitrogen (T-N) was, on average, 5.67–6.51 μM, ranging from 4.85 (station 2, Summer 2009) to 7.17 μM (station 1, summer 2010). NO_3^- and NH_4^+ concentrations were relatively high (> 1 μM), while NO_2^- concentration was much lower (0.22–0.24 μM) (**Table 1**). The concentration of total phosphorus (T-P) varied from 10.29 (station 1, summer 2009) to 15.97 μM (station 2, summer 2009). The relatively important T-P concentrations were due to the high contribution of PO_4^{3-}, close to 23% of T-P, which displayed a mean concentration of 3.25 ± 1.67 μM and 2.52 ± 0.42 μM in summer 2009 and summer 2010, respectively, showed the highest value 5.65 μM in summer 2009 station 2 (**Table 1**). Nutrient values were indicative of a generalized eutrophication. The N/P was always lower than the Redfield ratio, suggesting potential N limitation in this area (**Table 1**).

3.2 Zooplankton community structure and spatial distribution

The total zooplankton abundance varied from 2212 (station 2) to 65,688 ind m^{-3} (station 3) in summer 2010. Zooplankton assemblages were dominated by copepods,

DOI: http://dx.doi.org/10.5772/intechopen.1006102

Variables	**Summer 2009**			**Summer 2010**		
	Minimum	**Maximum**	**Mean ± SD**	**Minimum**	**Maximum**	**Mean ± SD**
Physical variables						
Temperature (°C)	27.74	28.56	28.20 ± 0.39	27.90	29.15	28.68 ± 0.54
Salinity (psu)	44.50	46.83	45.77 ± 0.96	44.70	46.30	45.46 ± 0.77
pH	8.21	8.61	8.35 ± 0.19	8.13	8.51	8.29 ± 0.16
Chemical variables						
NO_2^- (μM)	0.16	0.31	0.24 ± 0.07	0.11	0.39	0.22 ± 0.13
NO_3^- (μM)	0.99	1.28	1.11 ± 0.12	0.89	2.06	1.45 ± 0.58
NH_4^+ (μM)	1.09	1.46	1.23 ± 0.17	1.00	1.65	1.41 ± 0.31
T-N (μM)	4.85V	6.42	5.67 ± 0.81	6.12	7.17	6.51 ± 0.46
PO_4^{3-} (μM)	1.94	5.65	3.25 ± 1.67	2.03	3.05	2.52 ± 0.42
T-P (μM)	10.29	15.97	12.65 ± 2.76	11.82	12.54	12.22 ± 0.32
N/P ratio	0.50	1.23	0.92 ± 0.35	0.87	1.83	1.25 ± 0.42
$Si(OH)_4$ (μM)	2.09	4.25	3.51 ± 0.96	3.39	5.82	4.63 ± 1.09
Biological variables						
Chlorophyll-*a* (mg l^{-1})	0.83	1.15	1.00 ± 0.16	0.70	1.88	1.33 ± 0.54
T- zooplankton (ind m^{-3})	12,772	26,928	18,090 ± 6520	2212	65,688	40,810 ± 27,273
Copepods (ind m^{-3})	10,806	21,120	14,498 ± 4599	2170	62,470	39,076 ± 26,034
Cyclopoids (ind m^{-3})	1311	16,192	7032 ± 6419	917	247	12,862 ± 9709
Calanoids (ind m^{-3})	4224	8349	6392 ± 1705	501	24,704	11,893 ± 10,003

Variables	Summer 2009			Summer 2010		
	Minimum	Maximum	Mean ± SD	Minimum	Maximum	Mean ± SD
Harpacticoids (ind m^{-3})	000	310	118 ± 149	210	3360	1240 ± 1477
Poecilostomatoids (ind m^{-3})	000	176	119 ± 081	000	630	267 ± 263
Other zooplankton (ind m^{-3})	1478	5808	3592 ± 2187	042	3218	1734 ± 1320
T- ciliates (cells l^{-1})	000	700	081 ± 159	000	300	053 ± 088
T- phytoplankton (cells l^{-1})	1400	18,700	8450 ± 7424	9600	22,700	16,950 ± 6077

Table 1.
Minimum (Min), maximum (Max), and mean ± SD (standard deviation) of physicochemical parameters, zooplankton, phytoplankton, and ciliates communities in summer 2009/2010 along the El Bibane lagoon.

representing 73–89% and 95–98% of total zooplankton abundance in summer 2009 and 2010, respectively (**Figure 2**). Total copepod abundance was negatively associated with salinity ($r = -0.891$, $p > 0.05$) and positively with phosphates ($r = 0.923$, $p > 0.05$). The highest copepod abundances were observed at station 3 (62,470 ind m^{-3}) in July 2010. Their abundance did not exceed 21,120 ind m^{-3} in July 2009 (**Table 1**). Other zooplankton (Appendicularia, Bivalvia, *Cirripedia nauplii*, Cladoceran, Euphausiacea, Fish larvae, Gasteropoda larvae, Hydromedusae, Ostracoda, Polychaeta larvae and Zoea) presented low relative abundances at the two periods (2–26% of total zooplankton abundance) (**Figure 2**), with mean abundances of 1734 ± 1320 ind m^{-3} in summer 2010 and 3592 ± 2187 ind m^{-3} in summer 2009 (**Table 1**). Other zooplankton abundance was positively correlated with phosphates ($r = 0.985$, $p < 0.05$).

Copepods composition and abundance showed four groups: Calanoids (on average 19–77% of the total copepod abundance), Cyclopoids (12–76%), Harpacticoids (0–12%) and Poecilostomatoids (0–1%) (**Figure 3**). A total of 23 copepod species were found at all stations (**Table 2**), with *Oithona nana* dominating the total abundance of copepods (35%) in the summer of 2009. *Oithona similis* abundances were also higher in the summer of 2010 (104.73 ± 7279 ind m^{-3}) than in the summer of 2009 (2777 ± 497 ind m^{-3}). Copepod richness was higher in the summer of 2010 (21 species) than in the summer of 2009 (11 species) (**Table 2**). Among other

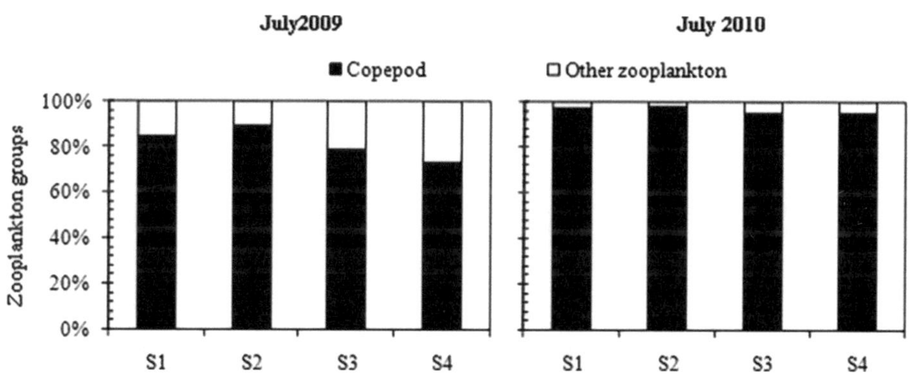

Figure 2.
Spatial variations of zooplankton groups abundance in summer 2009/2010 along the El Bibane lagoon.

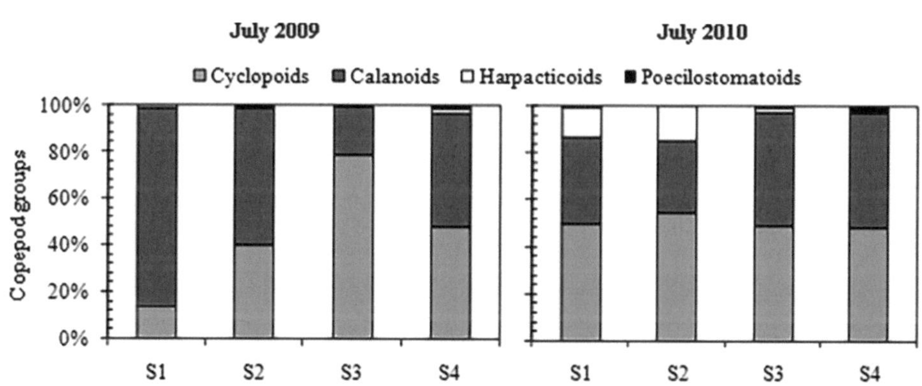

Figure 3.
Spatial variations of the relative abundance of copepod groups in summer 2009/2010 along the El Bibane lagoon.

	July 2009				July 2010			
	S1	S2	S3	S4	S1	S2	S3	S4
Copepoda								
Nauplii	982	591	528	1240	23,730	500	11,700	15,330
Calanoida								
Acartia clausi	328	—	—	—	8820	250	21,775	10,920
Acartia italica	—	—	—	—	—	—	98	—
Paracartia latisetosa	—	444	—	—	—	—	—	—
Centropages kroyeri	1474	148	352	620	—	—	231	—
Centropages typicus	4910	1035	2464	1705	315	—	325	—
Isias clavipes	—	—	—	—	210	—	325	210
Temora stylifera	—	—	—	—	210	—	—	—
Temora longicornus	164	—	—	—	—	42	—	210
Temora sp.	—	—	—	—	—	167	1950	210
Paracalanus parvus	1473	5169	1408	3875	—	42	—	1050
Megacalanus princeps	—	—	—	—	—	—	—	210
Cyclopoida								
Oithona nana	1147	2365	12,848	2635	420	417	6825	1890
Oithona similis	—	2217	3168	2945	12,600	500	17,875	10,920
Oithona plumifera	164	—	176	465	—	—	—	—
Poecilostomatoida								
Corycaeus clausi	—	—	—	—	—	—	—	210
Farranula rostrata	—	—	—	—	210	—	—	—
Oncaea conifera	—	148	176	155	—	—	—	420
Oncaea mediterranea	—	—	—	—	—	—	228	—
Harpacticoida								
Euterpina acutifrons	164	—	—	310	210	—	325	210
Microsetella norvegica	—	—	—	—	420	—	650	—
Microsetella rosea	—	—	—	—	1890	84	—	—
Macrosetella gracilis	—	—	—	—	—	84	—	—
Tisbe battagliai	—	—	—	—	840	84	163	—
Other Crustaceans								
Cladocera	328	148	4048	2790	210	—	325	—
Ostracoda	—	—	176	—	—	—	—	—
Euphausiacea	—	—	—	—	210	—	—	—
Gelatinous								
Appendicularia	328	739	704	930	—	—	—	—
Hydromedusae	—	—	—	—	315	42	325	—
Meroplankton								
Gasteropoda larvae	982	591	704	1085	210	—	650	210

	July 2009				July 2010			
	S1	S2	S3	S4	S1	S2	S3	S4
Bivalvia	328	—	176	310	—	—	293	210
Cirripedia nauplii	—	—	—	—	—	—	—	210
Polychaeta larvae	—	—	—	—	210	—	325	—
Zoea	—	—	—	—	210	—	—	420
Fish larvae	—	—	—	—	210	—	1300	1050

Table 2.
List and abundance (in ind.m^{-3}) of the zooplankton species observed in the summer 2009/2010 along the El Bibane lagoon.

zooplankton, Cladocerans were dominant in abundance (51% of total other zooplankton) in the summer of 2009, followed by Fish larvae with 37% in the summer of 2010 (**Figure 4**).

3.3 Relationships between zooplankton, phytoplankton, and ciliate

The spatial distribution of zooplankton abundance with the prevailing potential prey (total phytoplankton and total ciliate) is illustrated in **Figure 5**. The abundance of zooplankton does not show any significant correlations between the abundance of copepods or other zooplankton and phytoplankton and ciliates for both periods.

3.4 Multivariate analysis

The first factorial plane (axes 1 and 2) of the PCA analysis on environmental, phytoplankton, and ciliate variables explained 86.36% of the total variance,

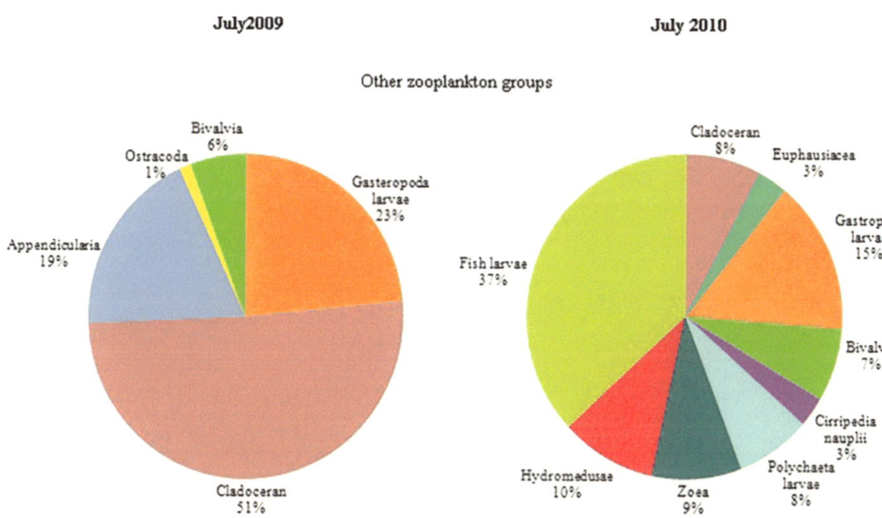

Figure 4.
Average distribution of the relative abundance of other zooplankton groups in summer 2009/2010 along the El Bibane lagoon.

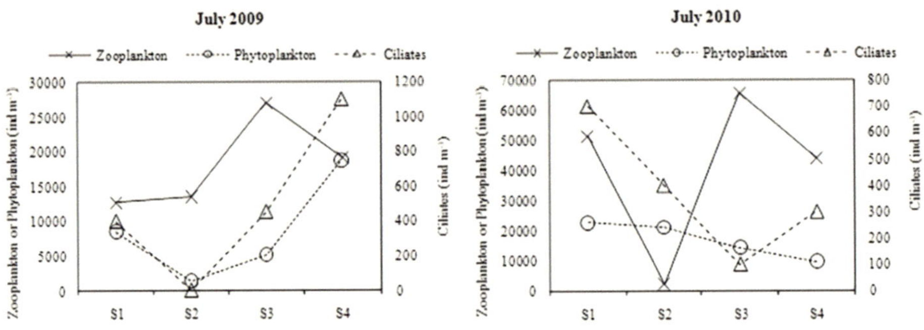

Figure 5.
Spatial variations of total zooplankton, phytoplankton, and ciliates abundance in summer 2009/2010 along the El Bibane lagoon.

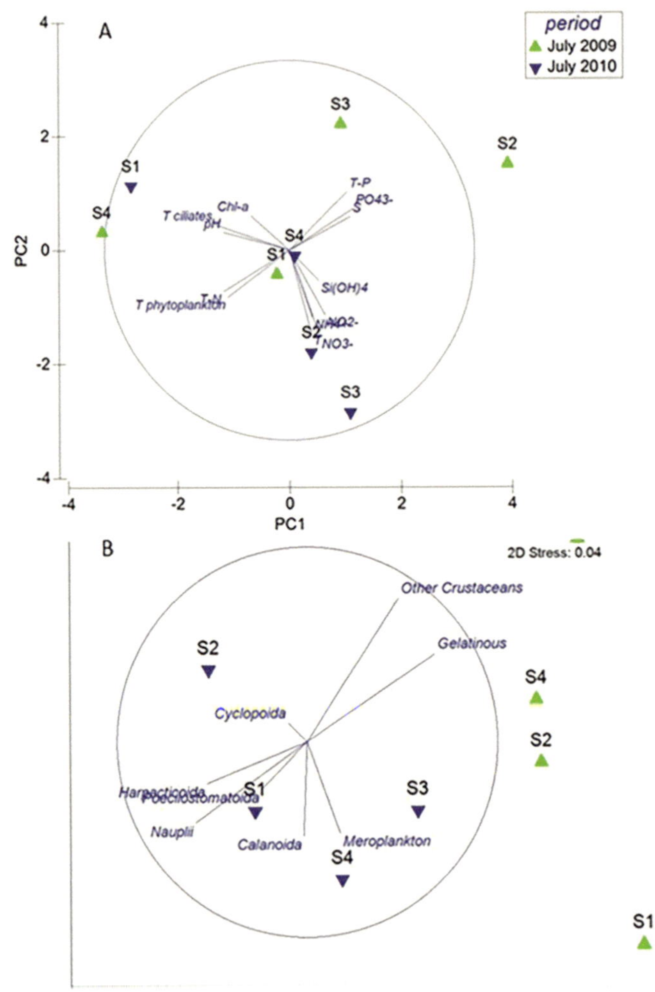

Figure 6.
Results of (A) principal component analysis (axes I and II) performed on environmental, phytoplankton, and ciliate variables and (B) non-metric multidimensional scaling on percentage abundance of zooplankton taxonomic groups.

58.35% of it for the first component and 28.01% for the second component (**Figure 6a**). The first axis shows an opposition between stations 2–3 correlated to all nutrients and stations 1–4 correlated to phytoplankton, Chl*a* pH, and ciliates. The second axis opposes the two periods for stations 2–3 associated with phosphorous nutrients in July 2009 and nitrogen nutrients and silicium in July 2010. The NMDS on zooplankton taxonomic groups clearly separates the communities of the two periods, with July 2009 associated with non-copepod crustaceans and gelatinous organisms and July 2010 associated with nauplii, Poecilostomatoide, Calanoida, and Harpacticoids.

4. Discussion

The present study is the first to examine the spatial distribution of zooplankton communities in the coastal waters of the El Bibane lagoon in relation to nutrients, phytoplankton, and ciliates. Nutrient concentrations showed a generalized eutrophication status (**Table 1**). High eutrophication levels of the coastal waters of the Gulf of Gabès were confirmed [13, 14].

Copepods dominated the zooplankton community in all the stations in the El Bibane lagoon in summer. The dominance of copepods has already been reported in several studies in the Gulf of Gabès region: on western area of the Djerba coasts (54–100% of total zooplankton abundances; [16]); on the northern coast of Sfax (61–82%; [2, 31]); along the southern coast of Kerkennah Islands (98%; [32]), in Kneiss Islands (30–96%; [33]) in summer. Among copepods, Calanoids were highly dominant (19–77% of the total copepod abundance), which is very similar to what was observed by Rekik et al. [16] in the coastal area of Djerba Island (up to 79% of total copepod abundance) and Drira et al. [34] on the coast of Sfax (43% of total copepod abundance).

Although the environmental and tropic conditions did not change significantly between the summer of 2009 and the summer of 2010, the structural composition of the zooplankton community showed very distinctive traits between the two periods in the El Bibane lagoon. Cladocerans were much more abundant in 2009, and appendicularians, relatively abundant in the four stations in 2009, were absent in 2010. The copepod community was dominated by *Oithona nana* (35% of total copepods) in the summer of 2009 and *Oithona similis* (40% of total copepods) in the summer of 2010. Small planktonic copepods reached important abundance throughout the study period. Small species, particularly Oithonids, were found to mostly dominate the copepod community in both summer 2009 and summer 2010. This appears to be a common feature in coastal areas of the Gulf of Gabès [2, 35] in offshore waters of the Gulf of Gabès [36]. The prominence of small planktonic copepods in the Gulf of Gabès, such as *Oithona nana* and *Oithona similis*, in diverse marine sites of the Gulf of Gabès, was inferred owing to its adaptive strategies [37], combined with an omnivorous diet [14] and lower metabolic needs [14], allowing it to tolerate environmental perturbations and tolerance to pollution [38]. The omnivorous/detritivorous diet can be confirmed by the lack of significant correlation between copepods and phytoplankton and between copepods and ciliates for both periods.

Oithona similis is known as the most omnipresent copepod species in the world sea [38]. This small species has a clear eurythermal and euryhaline distribution [39, 40].

5. Conclusion

This study analyzes zooplankton distribution in the El Bibane lagoon, emphasizing their connections with nutrients, phytoplankton, and ciliates. It confirms a generalized eutrophication status in the Gulf of Gabès and highlights the dominance of copepods, particularly Calanoids, consistent with regional findings. Significant changes in zooplankton composition between summer 2009 and summer 2010, despite stable conditions, are noted. The prominence of small planktonic copepods like *Oithona nana* and *Oithona similis* is linked to their adaptive strategies, including an omnivorous/detritivorous diet and tolerance to environmental disturbances. This study highlights the ecological importance of zooplankton as an environmental indicator and offers valuable insights into the marine ecosystems of the Gulf of Gabès. To address eutrophication, efforts in the Gulf focus on sustainable development, stricter industrial regulations, improved agricultural practices, and enhanced wastewater management.

Acknowledgements

This study was performed in the framework of the international project Société d'étude de Réalisation d'Aménagement et d'Hydraulique (SERAH). This work was conducted in the Marine Biodiversity and Environment LR/18ES30 laboratory at the University of Sfax.

Author details

Amira Rekik[1], Ahmad J. Al-Shemmari[2], Marc Pagno[3], Mohammad Ali[2*], Hanan Al-Adeelah[2], Adel Naseeb[4], Ahmad Al-Khayat[4], Mohammad Boarki[2], Neila Annabi-Trabelsi[1], Wassim Guermazi[1], Habib Ayadi[1] and Jannet Elloumi[1]

1 Faculty of Sciences of Sfax, Department of Sciences of Life, Laboratory LR/18 ES 30 Marine Biodiversity and Environment, University of Sfax, Sfax, Tunisia

2 Environment and Life Sciences Research Center, Kuwait Institute for Scientific Research, Safat, Kuwait

3 Aix Marseille University, CNRS/INSU, University of Toulon, IRD, Mediterranean Institute of Oceanography (MIO), Marseille, France

4 Techno-Economics Department, Kuwait Institute for Scientific Research, Safat, Kuwait

*Address all correspondence to: mohalikisr@gmail.com

References

[1] Alexander R. (diss.)Interactions of Zooplankton and Phytoplankton with Cyanobacteria. Lincoln, NE: University of Nebraska; 2012. 69 p

[2] Rekik A, Drira Z, Guermazi W, Elloumi J, Maalej S, Aleya L, et al. Impacts of an uncontrolled phosphogypsum dumpsite on summer distribution of phytoplankton, copepods and ciliates in relation to abiotic variables along the near-shore of the south western Mediterranean coast. Marine Pollution Bulletin. 2011;**64**:336-346

[3] Jacobsen S, Gaard E, Larsen KMH, Eliasen SK, Hátún H. Temporal and spatial variability of zooplankton on the Faroe shelf in spring 1997-2016. Journal of Marine Systems. 2018;**177**:28-38. DOI: 10.1016/j.jmarsys.2017.08.004

[4] Wickham SA. Trophic relations between cyclopoid copepods and ciliated protists: Complex interactions link the microbial and classic food webs. Limnology and Oceanography. 1995;**40**(6):1173-1181

[5] Dickey TD, Bidigare RR. Interdisciplinary oceanographic observations: The wave of the future. Scientia Marina. 2005;**69**:23-42

[6] Fernández De Puelles ML, Molinero JC. Decadal changes in hydrographic and ecological time-series in the Balearic Sea (western Mediterranean), identifying links between climate and zooplankton. ICES Journal of Marine Science. 2008;**65**:311-317

[7] Landry MR, Brown SL, Rii YM, Selph KE, Bidigare RR, Yang EJ, et al. Depth-stratified phytoplankton dynamics in cyclone opal, a subtropical mesoscale eddy. Deep-Sea Research Part. 2008;**II**(55):1348-1356

[8] Lin KY, Sastri AR, Gong GC, Hsieh CH. Copepod community growth rates in relation to body size, temperature, and food availability in the East China Sea: A test of metabolic theory of ecology. Biogeosciences. 2013;**10**:1877-1892

[9] Maar M, Rindorf A, FriisMøller E, Christensen A, Madsen KS, Van Deurs M. Zooplankton mortality in 3D ecosystem modelling considering variable spatial-temporal fish consumptions in the North Sea. Progress in Oceanography. 2014;**124**:78-91

[10] Leandro SM, Morgado F, Pereira F, Queiroga H. Temporal changes of abundance, biomass and production of copepod community in a shallow temperate estuary (Ria de Aveiro, Portugal). Estuarine, Coastal and Shelf Science. 2007;**74**:215-222

[11] Sherr E, Sherr B. Understanding roles of microbes in marine pelagic food webs: A brief history. In: Kirchman DL, editor. Microbial Ecology of the Oceans. 2nd ed. New York: John Wiley & Sons, Inc; 2009. pp. 27-44

[12] Stoecker DK, Michaels AE. Respiration, photosynthesis and carbon metabolism in planktonic ciliates. Marine Biology. 1991;**108**:441-447

[13] Annabi-Trabelsi N, Ali M, Belmonte G, Ayadi H, Guermazi W. Evaluation of the water quality and the eutrophication risk in Mediterranean Sea area: A case study of the Gulf of Gabès. In: Gonçalves AM, editor. Marine Ecosystems. Rijeka: IntechOpen; 2023. DOI: 10.5772/intechopen.114108

[14] Annabi-Trabelsi N, Guermazi W, Leignel V, Al-Enezi Y, Karam Q, Ali M, et al. Effects of eutrophication on plankton abundance and composition in the Gulf of Gabès (Mediterranean Sea, Tunisia). Water. 2022;**14**(14):2230

[15] Dolan JR, Coats DW. Seasonal abundances of planktonic ciliates and microflagellates in mesohaline Chesapeake Bay waters. Estuarine, Coastal and Shelf Science. 1990;**31**:157-175

[16] Rekik A, Guermazi W, Kmiha Megdiche S, Sellami I, Pagano M, Ayadi H, et al. Spatial variation of summer microphytoplankton and zooplankton communities related to environmental parameters in the coastal area of Djerba Island (Tunisia, Eastern Mediterranean). Mediterranean Marine Science. 2023;**24**(1):156-172. DOI: 10.12681/mms.30650

[17] Parsons TP, Maita Y, Lalli CM. A Manuel of Chemical and Biological Methods for Seawater Analysis. Oxford, England: Pergamon Press; 1984, 173 p

[18] Sherr EB, Sherr BF. Preservation and storage of samples for enumeration of heterotrophic protests. In: Kemp PF, Sherr BF, Sherr EB, Cole JJ, editors. Handbook of Methods in Aquatic Microbial Ecology. London: Lewis Publishers; 1993. pp. 207-212

[19] Grasshof KM. Determination of nitrate. In: Ehradt K, Grasshof KM, Kremling K, editors. Methods of Seawater Analyses. Verlag Chemie: Weinheim; 1983. pp. 143-150

[20] Utermöhl H. Zur Vervollkommung der quantitative Phytoplankton Methodik. Mitteilungen Internationale Vereinigungfür Theoretische und Angewandte. Limnologie. 1958;**9**:1-38

[21] Balech E. Tintinnoinea del Mediterraneo. Trabajos del Instituto Espanol de Oceanografia. 1959;**28**:1-88

[22] Tomas CR, Hasle GR, Steidinger AK, Syvertsen EE, Tangen C. Identifying Marine Diatoms and Dinoflagellates. San Diego, CA: Academic Press, Inc; 1996, 598 p

[23] Alder VA. Tintinnoinea. In: Boltovsky D, editor. South Atlantic Zooplankton. Leiden, The Netherlands: Backhuys Publishers; 1999. pp. 321-384

[24] Ciliophora PW. In: Boltovsky D, editor. South Atlantic Zooplankton. Leiden, The Netherlands: Backhuys Publishers; 1999. pp. 265-319

[25] Strüder-Kypke MC, Montagnes DJS. Development of web-based guides to planktonic protists. Aquatic Microbial Ecology. 2002;**27**:203-207

[26] Rose M. Copépodes pélagiques. In: Faune de la France. Vol. 26. Paris: Lechevalier; 1933, 368 p

[27] Costanzo G, Campolmi M, Zagami G. Stephosmarsalensis new species (Copepoda, Calanoida, Stephidae) from coastal waters of Sicily, Italy. Journal of Plankton Research. 2000;**22**:2007-2014

[28] Boxshall GA, Halsey SH. An Introduction to Copepod Diversity. Tome I. Printed and Bound by Henry Ling Ltd. Dorchester: The Dorset Press; 2003, 421 p

[29] Dolédec S, Chessel D. Rythmes saisonniers et composantes stationnelles en milieu aquatiqueII. Prise en compte et élimination d'effets dans un tableau faunistique. Acta Oecologica Oecologia Generalis. 1989;**10**:207-332

[30] Frontier S. Etude statistique de la dispersion du zooplancton. The Journal

of Experimental Marine Biology and Ecology. 1973;**12**:229-262

[31] Rekik A, Denis M, Aleya L, Maalej S, Ayadi H. Spring plankton community structure and distribution in the north and south coasts of Sfax (Tunisia) after north coast restoration. Marine Pollution Bulletin. 2013;**6**:82-93

[32] Rekik A, Ayadi H, Elloumi J. Distribution of the plankton assemblages during the winter and summer along the southern coast of the Kerkennah Islands (Tunisia, eastern Mediterranean Sea). Marine Ecology. 2018a;**39**(2):e12494

[33] Rekik A, Ayadi H, Elloumi J. Spatial and inter-annual variability of proto- and metazooplankton during summer around the Kneiss Islands (Tunisia, Central Mediterranean Sea). Applied Water Science. 2018b;**8**:99

[34] Drira Z, Kmiha-Megdiche S, Sahnoun H, Tedetti M, Pagano M, Ayadi H. Copepod assemblages as a bioindicator of environmental quality in three coastal areas under contrasted anthropogenic inputs (gulf of Gabes, Tunisia). Journal of the Marine Biological Association of the United Kingdom. 2018;**98**(8):1889-1905

[35] Ben Salem Z, Drira Z, Ayadi H. What factors drive the variations of phytoplankton, ciliate, and mesozooplankton communities in the polluted southern coast of Sfax, Tunisia? Environmental Science and Pollution Research. 2015;**22**(12):9071-9086. DOI: 10.1007/s11356-015-4416-8

[36] Ben Ltaief T, Drira Z, Hannachi I, Bel Hassen M, Hamza A, Pagano M, et al. What are the factors leading to the success of small planktonic copepods in the Gulf of Gabes, Tunisia? Journal of the Marine Biological Association of the United Kingdom. 2015;**95**:747-761

[37] Riccardi N, Mariotto L. Seasonal variations in copepod body length: A comparison between different species in the lagoon of Venice. Aquatic Ecology. 2000;**34**:243-252

[38] Gallienne CP, Robins DB. Is *Oithona* the most important copepod; in the world's oceans? Journal of Plankton Research. 2001;**23**:1421-1432

[39] Bel Hassen M, Drira Z, Hamza A, Ayadi H, Akrout F, Issaoui H. Summer phytoplankton pigments and community composition related to water mass properties in the Gulf of Gabes. Estuarine, Coastal and Shelf Science. 2008;**77**:645-656

[40] Annabi-Trabelsi N, Kobbi-Rebai R, Al-Enezi Y, Ali M, Subrahmanyam MNV, Belmonte G, et al. Factors affecting Oithona nana and Oithona similis along a salinity gradient. Mediterranean Marine Science. 2021;**22**(3):552

Section 3

Innovations and Applications

Chapter 7

Mathematical Modeling Is Unraveling the Metabolism of Photosynthetic Organisms to Drive Novel Culturing

Barbara Saucedo, Alexis Saldivar, Diego Martinez, Gabriela Canto-Encalada, Daniel Norena-Caro, Ila Peeler, Darshini Parekh, Jenna Armstrong, Mason Raymond, Shaila Prasad, Alexandra Guo and Cristal Zuniga

Abstract

Currently, our understanding of microalgae metabolism has been increasing due to the combination of experimental and computational tools developed for different kinds of photosynthetic microorganisms. The groundbreaking discoveries were enabled by high-throughput experiments and multi-scale modeling approaches for marine and freshwater microalgae, resulting in better understanding of single organisms and ecosystems. By connecting outcomes of the mathematical tools with big-scale datasets we are laying down the foundation for whole ecosystems modeling using environmentally relevant parameters. The revision of the state-of-the-art tools to understand the metabolism will facilitate and inform decisions for ecosystem restoration and production of commodities using light-driven biotechnology.

Keywords: microalgae, metabolism, genetic tools, metabolic models, artificial intelligence (AI), kinetic modeling

1. Introduction

All living organisms share an intricate network of chemical reactions that connects and sustains life which is referred to as metabolism. This process occurs within various biological scales from individual cells to full ecosystems on Earth. Between these scales are commonalities which reflect shared and organism-specific metabolic capabilities. Photosynthetic organisms offer an exciting platform for advancing both theoretical and applied metabolic research. We can enhance our understanding of physiology in regard to expression and metabolic pathways with metabolic models. These models allow us to develop testable hypotheses for practical applications,

IntechOpen

including ecosystem restoration and the production of valuable commodities. Recent developments in genetic tools, such as CRISPR/Cas9, RNA interference, and omics-based techniques, have significantly enhanced our ability to manipulate and optimize microalgal strains for industrial purposes. By modifying key metabolic pathways, scientists can increase the production of specific metabolites, such as lipids, pigments, and biofuels, while simultaneously improving stress tolerance and pollutant degradation. However, there is a challenge due to lack of tools and limitations in the broad applicability of genetic tools for non-conventional photosynthetic organisms. Both experimental and mathematical approaches face additional limitations such as regulation and low transformation rates. Despite the limitations, exploration of commonalities and differences will allow for approximation of effects and enable understanding of light-driven mechanisms.

To better understand metabolism at the microbial level scientists have developed computational pipelines and mathematical models. These models describe, test, calculate, and integrate metabolic processes through the establishment of balanced equations used to demonstrate biomass production rates of individual reactions, whole organisms, and ecosystems. A key advantage of using mathematical models across biological scales is the ability to integrate data from different levels of organization. To comprehend changes in chemical properties such as pH or temperatures (e.g. day and night cycles, climate change) we must be able to understand processes occurring in a single bacterium to untangle how the carbon cycle flows globally across-scale. By integrating information from single organisms to entire ecosystems, models become powerful tools for contextualizing large datasets and exploring the connections between life processes in response to environmental changes.

We envision that genome-scale metabolic modeling (GEM), artificial intelligence tools (AI) and high-throughput multi-omics technologies will facilitate the enrichment and organization of data in a meaningful way. In addition to this, questions should be answered about the effects of noisy environments and cellular stress on ecological timescales. With high-throughput experiments and model-driven approaches, scientists can quantify variables at previously unseen rates. Here we review the state-of-the-art of metabolic modeling, genetic tools, and AI tools applied to understand and modulate the metabolism of photosynthetic organisms.

2. Developed genetic tools for photosynthetic organisms and their applications for ecosystem restoration and specialty chemicals production

The main sources of anthropogenic water pollution, i.e., wastewater, landfill leachates, agricultural waste, and oil spills, contain different types of contaminants that do not inhibit, and even enhance blooms of photosynthetic organisms [1]. As a result, microalgae and cyanobacteria have traditionally been associated with detrimental impacts on the environment such as eutrophication of freshwater bodies and the oceanic red tides [2, 3]. Eutrophication is characterized by the uncontrolled abundance of nutrients. These experimental observations of high growth rates in natural environments opened a new field of light driven biotechnology using photosynthetic organisms as key players in bioproduction.

Interestingly, photosynthetic microorganisms have resulted in key participants in circular economy biotechnology, which aims to couple bioremediation operations with chemical production [4]. Specifically, microalgae encompass numerous potential applications because of their photosynthetic nature and ability to metabolize a wide range of chemical compounds with uncommon structures [5]. These natural capabilities, combined with advances in genetic engineering have profoundly transformed the study and application of marine and freshwater microalgae, especially within biotechnology and molecular biology. Innovative methods now enable the identification of industrially valuable microalgae and the enhancement of specific genetic traits. Through advancement screening techniques and cultivation methods, researchers can isolate promising organisms. Subsequently, molecular approaches, including omics-based tools and genetic engineering, can be employed to further refine and optimize these algae for various applications. Understanding metabolic pathways and genomes is essential for advancing genetic engineering, as they reveal the organization of metabolic pathways and identify critical regulatory elements that can optimize genetic modifications [6]. Therefore, the targeted overexpression of key enzymes can lead to an increased production of specific metabolites, like lipids, carbohydrates, or pigments [7].

2.1 High-throughput and genetic tools for marine and freshwater algae

Together with the big data era of the twenty-first century, new high-throughput technologies have been developed, resulting in new genetic and high-throughput experimental tools such as CRISPR/Cas9 and Adaptive Laboratory Evolution (ALE). Initially used in bacteria and fungi, these tools are also being applied to microalgae to improve traits such as growth rate, stress tolerance, and product yield by selecting for beneficial genetic mutations [8].

Furthermore, these experimental approaches enabled better understanding of the metabolism and have enhanced different microalgae species to attain industry potential. The use of genetic engineering in microalgae is not limited to a few species; a wide array of marine and freshwater microalgae has been subject to these techniques. **Table 1** summarizes the names of microalgae, application, and provides valuable insight into the breadth of research in this field. As expected, most of the tools have been optimized for *Chlamydomonas reinhardtii* (Chlorophyta) which is a model organism with growth capabilities over a broad range of conditions and organic substrates.

Development of genetic tools has enabled testing of regulatory elements, such as promoters and untranslated regions (UTRs), that regulate gene expression and stabilize transcripts. Additionally, codon optimization further enhances gene expression by aligning codon usage with the host's preferences. Selectable markers, like antibiotic and herbicide resistance genes, allow for the identification of successfully transformed cells, while screen-able markers such as GFP and LUC provide visual confirmation of gene expression. RNA interference (RNAi) enables gene silencing by using double-stranded RNA to specifically degrade target mRNAs and has been applied to various microalgae for functional studies [21]. Tools like zinc-finger nucleases (ZFNs), mega-nucleases (MNs), and transcription activator-like effector nucleases (TALENs) have been used for genome editing in microalgae, however CRISPR/Cas9 has become the most popular and advanced tool for photosynthetic organisms [22]. This is because CRISPR/Cas9 is simpler and more versatile compared to the other tools, which often face challenges such as high cost, complexity, or lower efficiency. The CRISPR-Cas9 system offers precise genome editing capabilities, allowing for targeted modifications that improve traits such as lipid production and

Genus	Species name	Type	Growth mode	Genetic tools	Application	Reference
Chlamydomonas	*Chlamydomonas reinhardtii*	Fresh	Photo-, hetero-, and mixotrophic	Gene editing (ZNF, TALEN, CRISPR), reporter genes, selectable markers, transformation	Pigments, biofuels, research model, wastewater treatment	[9]
Chlorella	*Chlorella vulgaris*	Fresh, Marine	Photo-, hetero-, and mixotrophic	Gene editing (CRISPR, ZFN, TALEN), RNAi, transformation	High production of proteins, polyunsaturated fatty acids, health supplements	[10]
Arthrospira / Spirulina	*Arthrospira platensis* (formerly *Spirulina platensis*)	Fresh, Brackish	Photo-, hetero-, and mixotrophic	Random mutagenesis methods, transformation	High production of proteins, nutritional supplements, biologics	[11]
Phaedactylaceae	*Phaeodactylum tricornutum*	Marine	Photoautotrophic	Gene editing (CRISPR), genome minimization, transformation	Pigments, biofuels, proteins	[12]
Botryococcus	*Botryococcus braunii*	Fresh, Brackish	Photo-, hetero-, and mixotrophic	Limited: genomic analysis, isolation, sequencing	Biofuels, hydrocarbons	[13]
Nannochloropsis	*Nannochloropsis oceanica*	Marine	Photoautotrophic	Gene editing (CRISPR), transformation, vectors	Fatty acid production, biofuels	[14]
Dunaliella	*Dunaliella salina*	Marine	Photo-, hetero-, and mixotrophic	CRISPR, transformation	Carotenoid production, cosmetics, pigments	[15]
Synechocystis	*Synechocystis pevalekii*	Fresh	Photoautotrophic	CRISPR, transformation	Biofuel, CO_2 fixation	[16]
Prochlorococcus	*Prochlorococcus marinus*	Marine	Photoautotrophic	Antibiotic resistance markers, CRISPR, GFP reporter genes, transformation	CO_2 fixation, bioremediation, model organism/research, climate change research/O_2 production	[17]
Chromochloris	*Chromochloris zofingiensis*	Fresh, Marine	Photo-, hetero-, and mixotrophic	CRISPR, mutagenesis, RNAi, transformation	Carotenoid production, pigments, lipid production, nutritional supplements, biofuel	[18]
Thalassiosira	*Thalassiosira pseudonana*	Marine	Photo-, mixotrophic	Gene Editing (CRISPR, RNAi, TALEN) transformation	Lipid production, biofuels, wastewater treatment, pharmaceuticals	[19]
Fistulifera	*Fistulifera solaris*	Marine	Photo-, mixotrophic	Genome sequencing, transformation	Biofuel, lipid production, bioremediation, pharmaceuticals/supplements, model organism	[20]

Table 1.
Summarizes various microalgae, highlighting their growth methods, genetic tools and applications.

biomass yield. Together, these advancements facilitate the development of microalgae with enhanced productivity and utility in industrial applications [21].

High-throughput phenotypic included during ALE experiments enable the study of genetic and expression trajectories of microorganisms when they are subjected to environmental stress conditions by culturing cells for hundreds or thousands of generations. ALE experiments can be performed manually or automatically. For photosynthetic organisms this approach has been applied to *Picochlorum* sp., a champion on bioenergy production, *Skeletonema costatuni*, a planktonic alga, and *C. reinhardtii* among other for acclimatation experiments [8]. During ALE experiments new strains can be enriched that can survive extreme conditions or various types of pollutants. In the next section we will review what those mechanisms are and their applications.

2.2 Mechanisms of pollution control with algae

Some microalgae have evolved specialized metabolic pathways to precipitate, assimilate, fixate, absorb, degrade, or accumulate harmful contaminants [23]. A significant group of contaminants are xenobiotics, which is a diverse group of natural and artificial emerging contaminants with atypical structures and complex degradation routes [23]. Currently, xenobiotics are consistently released to the environment in the form of pharmaceuticals, personal care products, agro-chemicals (such as fertilizers, herbicides, and pesticides), veterinary products, and industrial chemicals and their by-products [1]. While xenobiotics can have different chemical structures, the most common moieties comprise polycyclic aromatic hydrocarbons (PAH), halogenated organic compounds (e.g., dioxins and dioxin-like compounds), heavy metals (e.g., Lead, Mercury, Arsenic, Cadmium, and Chromium), steroids, and steroid derivatives [23, 24]. Considering these features, the underlying metabolic pathways supporting bioremediation with photosynthetic organisms involve enzymes that degrade phenolic compounds (i.e., polyphenol oxidase and laccase) [25, 26], broad-spectrum hydroxylation Cytochrome P450 monooxygenases (CYPs/P450s) [24], and exopolysaccharide mediated biosorption of heavy metals [27]. Although the degradation and assimilation of xenobiotic compounds require specialized metabolic pathways, essential nitrogen, phosphorus, and carbon nutrients are still necessary to maintain the base metabolism and balance oxidative stress, leading to increased activity levels of antioxidant enzymes, especially catalase, superoxide dismutase (SOD), glutathione reductase, and glutathione and ascorbate peroxidase [23]. While xenobiotics metabolism is gaining space in the design of bioremediation operations, synthetic biology and metabolic engineering strategies are also considering CYPs/P450s monooxygenases as bio-catalyzers for light-induced production of specialty chemicals in microalgae [28, 29]. Overall, the characterization of the metabolic capabilities of photosynthetic organisms has resulted in innovative methods to control pollution that in combination with heterotrophic organisms can be applied to restore ecosystems or for the development of bioremediation methods.

2.3 Algae-mediated bioremediation and photo-biocatalysis

Traditional methods of wastewater treatment, particularly those of physical and chemical background, often have cost and efficiency downsides [30]. Similarly, traditional remediation of polluted soil has been criticized for its low efficacy and further generation of pollutant byproducts [31]. Macroalgae and microalgae have been extensively reviewed as practical solutions for treating wastewater, soil, and air pollution [32]. Remarkably, green microalgae and cyanobacteria isolates have been shown to remove heavy metals from wastewater by ion biosorption, overcoming a hurdle in traditional physical techniques [33–35]. Effective applications for heavy-metal biosorption involve the formation of microalgal-bacterial consortia (MBC). Co-microalgal-bacterial consortia have been utilized as an alternative to traditional wastewater treatment. The consortium formed by activated sludge in tandem with a lab culture of *Chlorella vulgaris* (Chlorophyta) microalgae was effective for the treatment of municipal sewage at comparable rates to traditional filtration systems [36]. In another study, microalgal-microbial granular sludge reactors proved useful for removing carbon and ammonia from dairy wastewater [37]. Investigations on aromatic compounds degradation mechanisms point towards extracellular processes, especially in exopolysaccharide-producing cyanobacteria that have a positive effect on the growth of heterotrophic bacterial degraders [26]. For example, the consortium formed by *Pseudomonas* sp. (bacteria) and *Pseudanabaena* PP16 (cyanobacteria) isolated from pulp and paper wastewater was able to degrade between 70% and 100% of phenol in 24 h and ~80% of dichloroacetate after 1 week of treatment [25]. Another study on biological soil crusts reported that cyanobacteria-fire moss consortia were able to remedy soil erosion and restore soil fertility in post-fire ecosystems [38]. Similarly, Crouzet et al. [39] reported that cyanobacteria-soil algae communities can stabilize soil aggregates in cereal cropping systems found in dryland areas. While the biological treatment of air contaminants with microalgae has not been addressed in detail, Jo et al. studied the conditions to minimize the generation of irritant greenhouse gas nitrous oxide in algal-bacterial photobioreactors during wastewater treatment [40]. Similar efforts have been used to record and optimize the oxygen levels, and other culture conditions (pH, temperature, etc.) of cyanobacterial and bacterial-algal consortia for wastewater treatment [41, 42]. Recently, microalgae CYPs/P450s monooxygenases were reviewed to demonstrate their involvement in the biotransformation of pharmaceuticals and personal care products often found in aquatic environments. Interestingly, while green microalga *Chlamydomonas* sp. Tai-o3 could partially degrade ciprofloxacin and sulfadiazine antibiotics, *Chlorella* sp. could efficiently achieve nearly 97% biodegradation of thiamphenicol. In the same review, the CYPs/P450s monooxygenases of marine macroalgae, e.g., *Polysiphonia stricta* (formerly *Polysiphonia urceolata*), *Fucus vesiculosus* (Phaeophyceae), and *Ulva lactuca* (Chlorophyta), granted increased tolerance to xenobiotics, and the ability to serve as metabolic sinks for contaminants [43]. Currently, the striking ability of algae to degrade contaminants from the environment is serving as an inspiration to develop artificial photo-bio-catalyzers such as inorganic semiconductors, organic dyes, and natural materials to enhance cofactor regeneration and expand the degradation capabilities of light-driven bioremediation [44, 45].

2.4 Restoration facing global warming and acidification

Oceans are natural sinks of excess atmospheric carbon dioxide and seawater acidification is another detrimental consequence of rising greenhouse gas emissions [46]. Besides, higher temperatures and increasing levels of dissolved carbon dioxide in water bodies promote changes in the biodiversity of harmful and non-harmful phytoplankton (i.e., microalgae), especially on the surface ocean layer [47, 48]. Dinoflagellates, Chlorophytes, and cyanobacteria were identified as the most likely types of phytoplankton benefiting from warmer waters [49]. Among these, Dinoflagellates and cyanotoxin-producing cyanobacteria are under environmental scrutiny because they are responsible for harmful algal blooms (HAB), which might become more frequent in warmer non-tropical regions of the world [49, 50]. While enhanced harmful algae growth should be addressed as a serious environmental threat in the coming years, higher temperatures also open a window of opportunity for further biotechnological applications of thermophilic microalgae [51].

2.5 Microalgal commodities and high-value chemicals

In 2016, the global bioproduct economy was worth ~$280 billion, with an anticipated annual increase of 11% [52]. Harnessing the power of gene and metabolism editing, many studies have proposed genetic engineering in algae as a viable solution to further enhance the biotechnological production of chemicals [53, 54]. Currently, the applications of non-cyanotoxin-producing cyanobacteria and non-harmful Chlorophytes (green microalgae) for wastewater management and high-value chemical production are being enhanced by recent developments in microalgal in synthetic biology and metabolic engineering [29, 55, 56]. Microalgae are source of commodities (i.e., ethanol, butanol, fatty acids, organic acids) and specialty high-value chemicals (i.e., pigments, terpenes, secondary metabolites, and peptides) [5, 57, 58]. Algae of many species have been shown to produce secondary metabolites, which are important bioproducts in the highly valued biomedical sector, such as antibiotics, antivirals, antiparasitic and antifungal compounds [59]. Moreover, microalgae metabolites are dabbling in the world of novel cosmetics with prophylactic properties [60]. In the long list of high value bioproducts, microalgal pigments have become popular for their potential health benefits [61, 62]. In addition, several studies have cited the coculture of algae and bacteria, yielding biofuel, and even lignin, simultaneously [63–66]. With new technologies and monetary incentive, the optimization of bioproduction is a clear sequential step, which could also benefit from recent developments on Cytochrome P450-based biocatalysis for the production of novel drugs [28]. These emerging technologies can also utilize bacterial biosurfactants and offer a cost-effective alternative for purifying microalgae-based products, or even oil spill bioremediation efforts [67, 68], making light-driven processes more profitable. Given the broad range of metabolic properties of microalgae the development of computational methods is necessary. In the next section we will describe the reconstruction and simulation of kinetic and metabolic models, describing datasets and developed tools, that use AI and ML tools to unravel hidden properties of photosynthetic organisms.

3. Mathematical modeling of photosynthetic organisms

Each living cell manages a metabolic network consisting of metabolites, reactions, and genes. Elements and energy are constantly being channeled through this network in the form of metabolic fluxes, driving the production of macromolecules and biomass growth. These fluxes are influenced by the metabolic structure, environment and evolutionary adaptations. Additionally, mechanisms of replication and translation are also integral to understanding the metabolic network.

Mathematical models have been used to simulate nearly all biological processes at different levels. At metabolite level, genome-scale metabolic models are employed under steady-state conditions, while kinetic modeling is used in a dynamic framework. In both cases stoichiometric balances of supply and demand are described through the definition of equations and parameterization also referred to as constraints. Equations are built to describe specific circumstances and assumptions, such as the Michaelis-Menten equation or the Monod equation for enzymatic activities or whole cell growth in kinetic models and photoautotrophic or mixotrophic conditions in metabolic models. In this section basis and examples of kinetic and metabolic modeling are discussed in the context of their role in the enhancement of our knowledge about photosynthetic organisms.

3.1 Reconstruction, simulation and application of genome-scale metabolic models of algae and cyanobacteria

GEMs are a system biology tool that comprehensively represents the metabolic network of an organism (e.g. cyanobacteria, algae) [69]. Reconstruction of GEMs has significantly increased the efficiency of predicting species links among genotypes and phenotypes. These models are also effective for experimental design as well as to modulate functions and control of biological systems occurring within photosynthetic organisms. Additionally, GEMs give insights into cellular behavior under various nutritional conditions while optimizing metabolic pathways for biotechnological applications [69, 70]. GEMs are built based on genomic information that is used to extract all metabolic capabilities of a target organism by identifying proteins, metabolites, and reactions present and compiling them. GEMs describe the quantitative relationships between reactants and products for all the relevant biochemical reactions which is combined to create a mathematical matrix describing the stoichiometry of the entire metabolic network [71]. The flux through every reaction can then be simulated under various conditions using Flux Balance Analysis (FBA) [72]. GEMs are especially useful in studying the metabolism of a wide range of organisms, as they provide a holistic view of the interconnected pathways which govern cellular functions.

Metabolic models are also knowledge bases that are systematically curated using different computational methodologies. Model reconstruction consists of several steps, including retrieving the genomic and proteomic information of the target organism while identifying the basic metabolic structure of your microorganism of interest. Computational tools use this information to automatically reconstruct a draft model that it is subjected to manual curation to verify Gene-Reaction-Protein (GRP) associations, add a biomass objective function, and gap-fill pathways and reactions based on experimental data such as untargeted metabolomics or high-throughput phenotyping. Finally, addition of metadata to metabolites and reactions is critical to ensure compatibility and reusability in a sharable format such as JSON, MAT, SBML or XML [73].

Photosynthetic organisms present unique modeling challenges, such as the integration of light-dependent reactions with the broader metabolic network. As a result, GEMs have been generated for only a few well studied photosynthetic organisms (**Table A1**). *Chlamydomonas reinhardtii* is one of the most studied organisms, as shown by the frequency of updates that its model has had over the last decade (**Figure 1**).

The application of GEMs to photosynthetic organisms has led researchers to gain insights into how these organisms convert light energy into chemical energy and how this process interacts with other metabolic pathways [92]. The model of *C. reinhardtii* (iRC1080) was used to study lipid metabolism and light acclimatation, generating new hypotheses about the evolution of *C. reinhardtii* [75]. Similarly, the model of *Phaeodactylum tricornutum* (Bacillariophyceae),

DOI: http://dx.doi.org/10.5772/intechopen.1007463

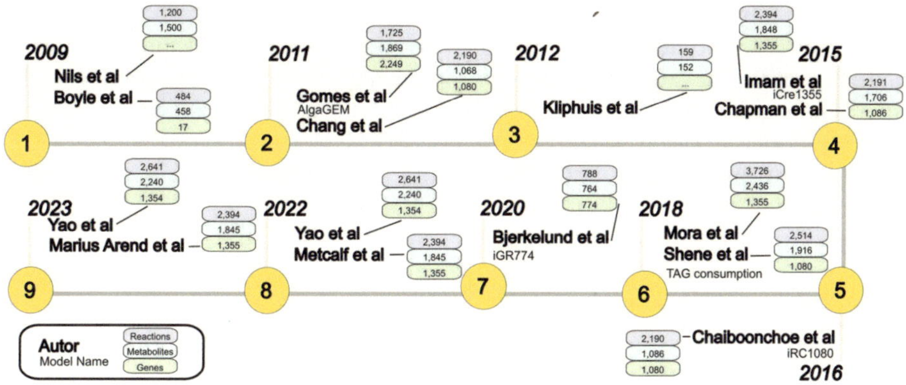

Figure 1.
Overview of the changes in the model properties of C. reinhardtii. *For biotechnology purposes, a model of* Auxenochlorella protothecoides *was curated to study several variables affecting lutein production at commercial levels, where sodium citrate was found to have the greatest effect on its production [83]. Additionally, the models of* Nannochloropsis oceanica *and salina, S3_iSO1949_Noceanica and iNS934, were used to design different media compositions which optimized triacylglycerol production [77, 79]. The model iAR632 of* Scenedesmus obliquus *was simulated in three growth modes (autotrophy, mixotrophy and heterotrophy) with mixotrophy being the one that favors the biomass and lipid production [78]. Similarly, the model of freshwater microalgae* Chlorella vulgaris *(iCZ843) simulates growth on photoautotrophic, heterotrophic, and mixotrophic modes with high accuracy [82], and a model of* Auxenochlorella pyrenoidosa *(formerly* Chlorella pyrenoidosa*) identified that cycling between autotrophic and heterotrophic modes was optimal for biomass production [84]. By combining GEMs and kinetic models, researchers can achieve a more comprehensive understanding of the metabolic processes in these organisms, which is crucial for applications in bioengineering and sustainable energy production.*

iLB1027_lipid, was used to study the relationship between the biochemical composition of biomass and the dynamic equilibrium between photosynthesis and heterotrophic metabolism during light-dark cycles [74]. The model of *Thalassiosira pseudonana* (Mediophyceae), CCMP 1335, was created with the objective to investigate the dissipation of reductants generated through light-dependent processes and the role of nitrate and sulfate assimilation to predict the availability of these compounds in the ocean surface [80]. The model of *Microchloropsis gaditana* (formerly *Nannochloropsis gaditana*) (Eustigmatophyceae), iRJ1321, was used to study the effect of nitrogen limitation on the Calvin cycle [81]. Meanwhile, the model of *Cylindrotheca* (Bacillariophyceae), iMK1961, was used to identify the role of this microorganism in global carbon cycling [70].

3.2 Basis and applications of kinetic models of algae and cyanobacteria

Kinetic models are used to predict microalgae growth by analyzing the relationship between environmental inputs (such as nutrients, light, and temperature) and outputs like biomass production. Two main approaches to the kinetic models exist: one focuses on optimizing external growth conditions to maximize yields, while the other examines how carbon is distributed within the cells, however both are important for optimizing growth and productivity [21]. Unlike GEMs, which are typically constraint-based and operate under steady-state assumptions, kinetic models incorporate the rates of the biochemical reactions, allowing for the simulation of time-dependent changes in metabolite concentration [93].

The generation of kinetic models requires a more detailed understanding of enzyme kinetics and reaction mechanisms [94] and they are often more computationally intensive and parameter-fitting-demanding than GEMs due to the need for extensive kinetic parameters. The first step involves gathering parameters, such as reaction rates and substrate affinities, usually derived from complex in vitro experiments [95]. This data is then used to parameterize rate laws, such as Michaelis-Menten kinetics, for each enzymatic reaction [95]. Once the kinetic parameters are defined, the model is calibrated using experimental data to ensure it accurately represents the dynamics of the system under various conditions [95].

Kinetic models are also used to quantify the effects of nutrient availability and physicochemical conditions. These models can represent only one of these limiting factors at a time using a threshold (also called minim law), or multiple factors using a multiplicative law [96]. The effect of light availability on growth rate can be described by saturation kinetics such as the Monod equation, although several modifications to consider the effect of photo-inhibition have been proposed [97, 98]. Alternatively, the effects of photo-inhibition can be quantified by using semi-structured models in which biomass is segregated into active, resting and inhibited fractions [99, 100]. Segregated models are also capable of quantifying the effects of nutrient limitations [101]. Because of absorbance and scattering effects of cell particles, light availability is heterogeneous inside of photo-bioreactors. The heterogeneity can be modeled using the Beer-Lambert equation, which is a good compromise between accuracy and computational costs, although quantifying the effects of light scattering is possible using the laws of radiative transfer [102–104]. Finally, multi-level models can integrate metabolism, light transfer and fluid dynamics, increasing the descriptive capabilities at the cost of computational complexity [99].

4. Large-scale resources available of photosynthetic organisms can enable better artificial intelligence and machine learning approaches

4.1 Large-scale data generates environmentally relevant parameters and constraints

Biogeochemical cycles have co-evolved over hundreds of years together with metabolic pathways. They are the biggest biological system altering chemistry in the ocean, atmosphere, and terrestrial ecosystems [105]. Those cycles have been studied for long periods of time through the collection of samples and their associated properties (metadata). Resources such as the Earth Microbiome Project (EMP), the National Microbiome Data Collaborative (NMDC) from DOE and the National Ecological Observatory Network (NEON) from NSF contain large-scale data publicly available for academic purposes [106–108]. The EMP dataset focused on sequencing the 16S rRNA gene to profile the taxonomic composition of 23,828 samples from 97 different studies. Meanwhile, the NMDC dataset has whole genome sequencing data for 4333 samples. Data mining plays a critical role during the development of new computational tools through the assessment of multi-omics data (e.g. metagenome, transcriptome, metabolome, and proteome) in EMP, NMDC, and NEON. Model-driving findings can be compared with experimentally observed patterns in natural systems.

In the work of Pushpakumara et al. [109] the authors used 10 studies in the EMP dataset to identify key bacterial associates in microalgae associated microbiomes. In particular, the authors used sequences from the 16S rRNA gene found in the chloroplast to infer co-occurrence patterns between microalgae and bacteria. They found that members of the phyla Bacteroidetes, Proteobacteria, Planctomycetes and

Verrucomicrobia are tightly associated with microalgae taxa [109]. Our analysis of the metadata of the EMP dataset (filtering out sequences originated in the chloroplast) shows that Cyanobacteria were highly abundant across ecosystems, present in 68% of all samples (23,828 total). **Figure 2** shows how abundance ratios highly change when dividing the samples in six groups for different temperature and pH ranges. Metadata collected from these studies showed that pH and temperature greatly influence the abundance of organisms such as Bacillus in all collected samples (**Figure 2**). On the contrary, Cyanobacteria genera such as *Synechococcus* and *Phormidium* remain highly abundant among five of the different groups of samples, demonstrating the capacity of these organisms to grow at non-optimal conditions. This meta-analysis highlights the physiologic robustness of Cyanobacteria, a property that is desirable to preserve the stability of production processes, especially when cultivated in outdoor conditions. Analysis of data collected directly from ecosystems will facilitate hypothesis testing and increase the likelihood of success for the development of interventions.

4.2 Genome analysis and gene localization

Modeling for photosynthetic organisms has the added complexity of determining compartmentalization. Eukaryotic microalgae contain different organelles, and the exact location a protein can be found is species dependent [111]. The same reaction may occur in two different compartments but be facilitated by different proteins. This

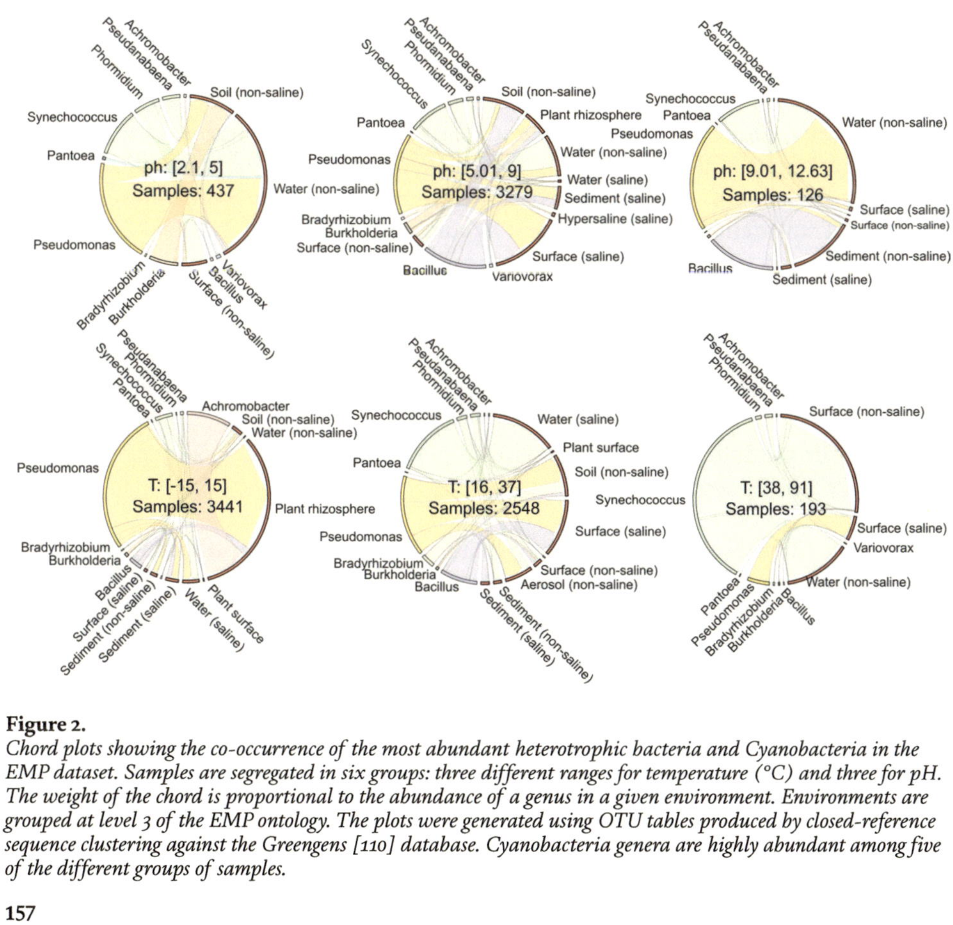

Figure 2.
Chord plots showing the co-occurrence of the most abundant heterotrophic bacteria and Cyanobacteria in the EMP dataset. Samples are segregated in six groups: three different ranges for temperature (°C) and three for pH. The weight of the chord is proportional to the abundance of a genus in a given environment. Environments are grouped at level 3 of the EMP ontology. The plots were generated using OTU tables produced by closed-reference sequence clustering against the Greengens [110] database. Cyanobacteria genera are highly abundant among five of the different groups of samples.

is why it is important to consider where each enzyme/protein is localized in the cell. Accurate annotation of proteins and compartmentalization in the model is necessary for maximizing information content and gaining detailed knowledge about microalgal metabolism [112]. A metabolic model's predictive capabilities is highly dependent on the accurate representation of metabolic exchange between organelles. Proteomics techniques can be used to experimentally identify localization of proteins in an organism [113]. However, these experimental techniques have their limitations and are difficult to implement for an entire genome [113]. AI and machine learning tools can be used to determine the localization of each protein within a genome and increase the quality of the metabolic model.

Machine learning tools allow for an increase of understanding of gene functions that can be expanded to genome wide. One tool, PredAlgo, has been developed specifically aimed at microalga predictions. PredAlgo uses a neural network machine learning technique to assign localization predictions to algal proteins [114]. The training set was developed from *C. reinhardtii* experimental data. The tool is capable of classifying proteins into three compartments: Chloroplast (C), mitochondrion (M) and secretory pathway (SP) [114]. PredAlgo calculates a score for the three cellular compartments and assigns a classification based on set cutoffs. If the three scores are below a certain cutoff, the protein was assigned to the "Other" (O) category [114]. This tool has been used in previous studies like lipid degradation [115] and identifying key proteins in *C. reinhardtii* needed during photosynthesis [116]. While PredAlgo is specifically for green algae, other tools have been developed to analyze and predict localizations of eukaryotic organisms.

DeepLoc is a neural network machine learning tool that can predict the subcellular localization of eukaryotic proteins [117]. This tool can differentiate between 10 different protein classes: Nucleus, Cytoplasm, Extracellular, Mitochondrion, Cell membrane, Endoplasmic reticulum, Chloroplast, Golgi apparatus, Lysosome/Vacuole and Peroxisome [117]. The latest version, DeepLoc 2.1, additionally classifies membrane association of a protein [118]. The membrane association can be defined as peripheral, transmembrane, lipid anchor or soluble. This algorithm allows for multi-label identifiers in which a protein can be localized into multiple compartments. It has benefited research including a study that analyzed the viral protein interactions and localizations of three coronaviruses, including SARS-CoV-2 [119]. With the continued advancement in AI and machine learning tools, protein annotations and localizations can be determined and further improved for genomes of photosynthetic organisms.

4.3 Metabolic modeling and artificial intelligence enhance our understanding of photosynthetic organisms

The high-throughput generation of genome sequences has significantly enhanced the understanding of gene functionality in organisms [120]. Genes encode proteins that perform diverse roles, including catalyzing metabolic reactions essential for cellular processes. One of the most widely used tools for studying metabolic networks in photosynthetic organisms is the flux balance analysis (FBA) [111]. FBA is a mathematical approach that calculates metabolic fluxes while optimizing an objective function (usually biomass production). Although FBA is a very useful tool for elucidating metabolism, it is still constrained to a single objective function [121]. Therefore, different efforts were applied to develop an extended FBA analysis in photosynthetic organisms that optimizes multi-objective functions [121]. Briones-Baez et al. [121] proposed an evolutionary algorithm capable of optimizing multiple objective functions by getting a good approximation of the Pareto frontier (**Figure 3**). The Pareto frontier represents a set of different optimal solutions where no single objective can be improved without worsening another. Therefore, the new algorithm was able to find a set of optimal solutions that balance trade-offs between two or more competing objectives. The algorithm was tested on the microalgae *C. reinhardtii*. They could successfully optimize simultaneously three metabolic objectives (protein production, carbohydrate production, and CO_2 uptake).

A more advanced method for studying the physiological functions of photosynthetic organisms is through the application of digital twins [122]. A digital twin is a virtual representation of a real-world physical asset, system, or process by integrating real-time data from sensors, instruments, and mathematical models [122]. This innovative technique, if applied in photosynthetic systems, can be used to predict ecosystem and climate changes and to test hypotheses in a virtual environment [122]. For example, the generation of a plankton digital twin can be used to simulate the ecological interaction among plankton species in their natural habitats [122]. Since the plankton population is a good indicator of climate change, researchers can predict long-term impacts on the environment and develop mitigation strategies [122].

The integration of genetic engineering tools and Machine Learning (ML) techniques in microalgae processes has allowed a better understanding of microalgae metabolism and genetic information. In combination with next-generation sequencing techniques, ML methods improved the acquisition of a wide range of genome sequence features without requiring full sequencing details [123] . For example, different studies showed that ML methods helped with the identification of differentially expressed genes, the detection of meta-genes related to a specific microalga pathway, among others [123]. Furthermore, artificial intelligence (AI) algorithms can also be applied to the design of experiments that involve microalgae, such as the determination of microalgae concentration, microalgae characterization, microalgae conversion, microalgae cultivation and species, and strain identification [123]. For example, Long et al. [124] utilized ML models to predict the light distribution patterns in algal cultures, which allowed the optimization of a semi-continuous bioreactor for algal cultivation.

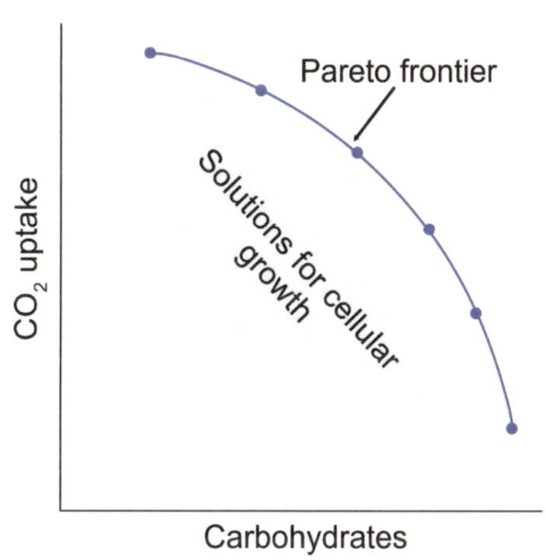

Figure 3.
The Pareto frontier shows the trade-off between CO_2 uptake and carbohydrates. Different solutions that allow cellular growth are represented in the shadow area, while the optimal trade-offs are highlighted with the blue line.

The applications of microalgae for bioproduct production are not limited to low scale. Different efforts have studied the use of microalgae in wastewater treatment processes as a sustainable alternative [125]. However, the synchronicity of choosing the correct algal strain, optimum temperature, pH, type of bioreactor, CO_2, light intensity, and media composition is a complex challenge for achieving optimal growth conditions [125]. Thus, ML algorithms have been developed to find the best conditions for enhancing biomass production at larger scales. Ansari et al. [126] utilized a three-layer feed-forward back-propagation artificial neural network (ANN) model to predict the algal dry cell weight in an outdoor pilot-scale system using circular pools as a cultivation system. The parameters used as inputs for the model were temperature, pH, dissolved oxygen (DO), electrical conductivity (EC), nitrate (NO_3^-), and phosphate (PO_4^{3-}), being NO_3^- and PO_4^- the most influential factors.

In recent years, microalgae have gained recognition as a sustainable alternative for producing value-added products. Consequently, advanced technologies for understanding and applying microalgae at an industrial scale have been developed. AI tools and ML algorithms are set to play a crucial role in optimizing these bioprocesses.

Acknowledgements

This work was supported by the National Science Foundation (Grant No. 2313313), the Department of Energy (DE-SC0025445) and the United States Department of Agriculture (USDA-NIFA-AFRI 2024-67017-42676). Any opinions, findings, and conclusions or recommendations expressed in this material are those of the author(s) and do not necessarily reflect the views of the National Science Foundation.

Appendix

ID	Name	Metabolites	Reactions	Genes	Reference
iLB1027_lipid	*Phaeodactylum tricornutum* CCAP 1055/1	2172	4456	1027	[74]
iRC1080	*Chlamydomonas reinhardtii*	1706	2191	1086	[75]
iCY1170_DHA	*Schizochytrium limacinum* (Thraustochytrids) SR21	1659	1769	1170	[76]
S3_iSO1949_Noceanica	*Nannochloropsis oceanica*	3018	3485	1981	[77]
iAR632	*Tetradesmus obliquus* (formerly *Scenedesmus obliquus*)	734	1467	632	[78]
iNS934	*Microchloropsis salina* (formerly *Nannochloropsis salina*)	1985	2345	934	[79]
CCMP 1335	*Thalassiosira pseudonana*	2,792	6,079	1,432	[80]
iRJ1321	*Microchloropsis gaditana* (formerly *Nannochloropsis gaditana*)	1862	1918	1321	[81]
iCZ843	*Chlorella vulgaris*	1770	2294	843	[82]
	Auxenochlorella protothecoides	2121	1974	0	[83]

ID	Name	Metabolites	Reactions	Genes	Reference
	Auxenochlorella pyrenoidosa (formerly *Chlorella pyrenoidosa*)	61	67	0	[84]
	Tetradesmus obliquus	183	351	0	[85]
iMK1961	*Cylindrotheca closterium*	3559	6718	1961	[70]
iJN678	*Synechocystis* sp. PCC6803	795	863	622	[86]
iJB785	*Synechococcus elongatus* PCC7942	768	850	785	[87]
iMS837	*Synechococcus elongatus* PCC7942	801	887	837	[88]
iAK888	*Arthrospira platensis* C1	1096	994	888	[89]
	Arthrospira platensis NIES-39	673	746	620	[90]
iCyj826	*Cyanothece* sp. PCC 7822	1110	1258	826	[91]

Table A1.
Overview of the genome-scale metabolic models available for photosynthetic organisms.

Author details

Barbara Saucedo[1†], Alexis Saldivar[1,2†], Diego Martinez[1], Gabriela Canto-Encalada[1,2], Daniel Norena-Caro[1], Ila Peeler[1,2], Darshini Parekh[1,3], Jenna Armstrong[1,3], Mason Raymond[1], Shaila Prasad[1], Alexandra Guo[1] and Cristal Zuniga[1*]

1 Department of Biology, San Diego State University, San Diego, CA, USA

2 Cellular and Molecular Biology Joint Doctoral Program with UC San Diego, San Diego State University, San Diego, CA, USA

3 Bioinformatics and Medical Informatics Graduate Program, San Diego State University, San Diego, CA, USA

*Address all correspondence to: czuniga2@sdsu.edu

†These authors contributed equally to this work.

References

[1] Kaur S, Reddersen B. Algae based solutions for polluted environments to restore ecosphere equilibrium. International Journal of Environmental Pollution and Remediation (IJEPR). 2022;**10**:9-18

[2] Ma J et al. Cyanobacteria in eutrophic waters benefit from rising atmospheric CO_2 concentrations. Science of the Total Environment. 2019;**691**:1144-1154

[3] Pitcher GC, Probyn TA. Suffocating phytoplankton, suffocating waters—Red tides and anoxia. Frontiers in Marine Science. 2016;**3**

[4] Chhandama MVL, Rai PK, Lalawmpuii. Coupling bioremediation and biorefinery prospects of microalgae for circular economy. Bioresource Technology Reports. 2023;**22**:101479

[5] Noreña-Caro D, Benton MG. Cyanobacteria as photoautotrophic biofactories of high-value chemicals. Journal of CO2 Utilization. 2018;**28**:335-366

[6] Qin S, Lin H, Jiang P. Advances in genetic engineering of marine algae. Biotechnology Advances. 2012;**30**(6):1602-1613

[7] Radakovits R et al. Genetic engineering of algae for enhanced biofuel production. (1535-9786 (Electronic))

[8] LaPanse AJ, Krishnan A, Posewitz MC. Adaptive laboratory evolution for algal strain improvement: Methodologies and applications. Algal Research. 2021;**53**:102122

[9] Kong F et al. Molecular genetic tools and emerging synthetic biology strategies to increase cellular oil content in *Chlamydomonas reinhardtii*. Plant and Cell Physiology. 2019;**60**(6):1184-1196

[10] Ru ITK et al. Chlorella vulgaris: A perspective on its potential for combining high biomass with high value bioproducts. Applied Phycology. 2020;**1**(1):2-11

[11] Jester BW et al. Development of spirulina for the manufacture and oral delivery of protein therapeutics. Nature Biotechnology. 2022;**40**(6):956-964

[12] Pampuch M, Walker EJL, Karas BJ. Towards synthetic diatoms: The Phaeodactylum tricornutum Pt-syn 1.0 project. Current Opinion in Green and Sustainable Chemistry. 2022;**35**:100611

[13] Boland DJ et al. Reclassification of *Botryococcus braunii* chemical races into separate species based on a comparative genomics analysis. PLoS One. 2024;**19**(7):e0304144

[14] Poliner E et al. A toolkit for *Nannochloropsis oceanica* CCMP1779 enables gene stacking and genetic engineering of the eicosapentaenoic acid pathway for enhanced long-chain polyunsaturated fatty acid production. Plant Biotechnology Journal. 2018;**16**(1):298-309

[15] Hu L et al. CRISPR/Cas9-induced β-carotene hydroxylase mutation in *Dunaliella salina* CCAP19/18. AMB Express. 2021;**11**(1):83

[16] Liu D, Pakrasi HB. Exploring native genetic elements as plug-in tools for synthetic biology in the cyanobacterium *Synechocystis* sp. PCC 6803. Microbial Cell Factories. 2018;**17**(1):48

[17] Li J et al. Clustered regularly interspaced short palindromic repeat/

CRISPR-associated protein and its utility all at sea: Status, challenges, and prospects. Microorganisms. 2024;**12**(1):118

[18] Koren I et al. Chromochloris zofingiensis (Chlorophyceae) divides by consecutive multiple fission cell-cycle under batch and continuous cultivation. Biology (Basel). 2021;**10**(2):157

[19] Nawaly H, Tsuji Y, Matsuda Y. Rapid and precise genome editing in a marine diatom, *Thalassiosira pseudonana* by Cas9 nickase (D10A). Algal Research. 2020;**47**:101855

[20] Naser I et al. Highly efficient genetic transformation methods for the marine oleaginous diatom *Fistulifera solaris*. Marine Biotechnology. 2023;**25**(5):657-665

[21] Sun H et al. Powerful tools for productivity improvements in microalgal production. Renewable and Sustainable Energy Reviews. 2021;**152**:111609

[22] Kumar G et al. Bioengineering of microalgae: Recent advances, perspectives, and regulatory challenges for industrial application. Frontiers in Bioengineering and Biotechnology. 2020;**8**

[23] Priya AK et al. Recent advances in microbial-assisted degradation and remediation of xenobiotic contaminants; challenges and future prospects. Journal of Water Process Engineering. 2024;**60**:105106

[24] Schröder GC, Smit MS, Opperman DJ. Harnessing heme chemistry: Recent advances in the biocatalytic applications of cytochrome P450 monooxgenases. Current Opinion in Green and Sustainable Chemistry. 2023;**39**:100734

[25] Kirkwood AE, Nalewajko C, Fulthorpe RR. The effects of cyanobacterial exudates on bacterial growth and biodegradation of organic contaminants. Microbial Ecology. 2006;**51**(1):4-12

[26] Żyszka-Haberecht B, Niemczyk E, Lipok J. Metabolic relation of cyanobacteria to aromatic compounds. Applied Microbiology and Biotechnology. 2019;**103**(3):1167-1178

[27] Chugh M et al. Algal bioremediation of heavy metals: An insight into removal mechanisms, recovery of by-products, challenges, and future opportunities. Energy Nexus. 2022;**7**:100129

[28] Wei Y, Ang EL, Zhao H. Recent developments in the application of P450 based biocatalysts. Current Opinion in Chemical Biology. 2018;**43**:1-7

[29] Urlacher VB, Girhard M. Cytochrome P450 monooxygenases in biotechnology and synthetic biology. Trends in Biotechnology. 2019;**37**(8):882-897

[30] Crini G, Lichtfouse E. Advantages and disadvantages of techniques used for wastewater treatment. Environmental Chemistry Letters. 2019;**17**(1):145-155

[31] Valdiviezo Gonzales LG et al. Scientometric study of treatment technologies of soil pollution: Present and future challenges. Applied Soil Ecology. 2023;**182**:104695

[32] Shahi Khalaf Ansar B et al. Removal of organic and inorganic contaminants from the air, soil, and water by algae. Environmental Science and Pollution Research. 2023;**30**(55):116538-116566

[33] Heidarpour A et al. Bio-removal of Zn from contaminated water by using green algae isolates. Environmental

Technology & Innovation. 2019;**16**:100464

[34] Kamcev J et al. Salt concentration dependence of ionic conductivity in ion exchange membranes. Journal of Membrane Science. 2018;**547**:123-133

[35] Ubando AT et al. Microalgal biosorption of heavy metals: A comprehensive bibliometric review. (1873-3336 (Electronic))

[36] Arango J et al. The potential of membrane-assisted microalgae-bacteria consortia for the treatment of real municipal sewage. Journal of Water Process Engineering. 2023;**56**:104527

[37] Bucci P et al. Assessment of the performance of a symbiotic microalgal-bacterial granular sludge reactor for the removal of nitrogen and organic carbon from dairy wastewater. Chemosphere. 2024;**351**:141250

[38] Muñoz-Rojas M et al. Restoring post-fire ecosystems with biocrusts: Living, photosynthetic soil surfaces. Current Opinion in Environmental Science & Health. 2021;**23**:100273

[39] Crouzet O, Consentino L, Pétraud JP, Marrauld C, Aguer JP, Bureau S, et al. Soil photosynthetic microbial communities mediate aggregate stability: Influence of cropping systems and herbicide use in an agricultural soil. Frontiers in Microbiology. 2019;**10**:1319

[40] Jo Y et al. Assessing nitrous oxide emissions from algal-bacterial photobioreactors devoted to biogas upgrading and digestate treatment. Chemosphere. 2024;**361**:142528

[41] Zhao W, Zhang G. Optimization of photosynthetic bacteria wastewater treatment and study of microbial species diversity. Desalination and Water Treatment. 2014;**52**(28-30):5357-5365

[42] Sánchez Zurano A et al. Modeling of photosynthesis and respiration rate for microalgae–bacteria consortia. Biotechnology and Bioengineering. 2021;**118**(2):952-962

[43] Li S et al. Cytochrome P450 enzyme-based biotransformation of pharmaceuticals and personal care products (PPCPs) by microalgae in the aquatic environment. Chemical Engineering Journal. 2023;**476**:146557

[44] Meyer L-E, Eser BE, Kara S. Coupling light with biocatalysis for sustainable synthesis—Very recent developments and future perspectives. Current Opinion in Green and Sustainable Chemistry. 2021;**31**:100496

[45] Peng Y et al. Recent advances in photobiocatalysis for selective organic synthesis. Organic Process Research & Development. 2022;**26**(7):1900-1913

[46] Doney SC et al. Ocean acidification: The other CO_2 problem. Annual Review of Marine Science. 2009;**1**:169-192

[47] Landry MR et al. Phytoplankton growth and grazing dynamics during anomalous heat wave and suppressed upwelling conditions in the southern California current. Deep Sea Research Part I: Oceanographic Research Papers. 2024;**210**:104353

[48] Gobler CJ. Climate change and harmful algal blooms: Insights and perspective. Harmful Algae. 2020;**91**:101731

[49] Brandenburg KM, Velthuis M, Van de Waal DB. Meta-analysis reveals enhanced growth of marine harmful algae from temperate regions with warming and elevated CO_2 levels. Global Change Biology. 2019;**25**(8):2607-2618

[50] Igwaran A et al. Cyanobacteria harmful algae blooms: Causes, impacts, and risk management. Water, Air, & Soil Pollution. 2024;**235**(1):71

[51] Patel A et al. A perspective on biotechnological applications of thermophilic microalgae and cyanobacteria. Bioresource Technology. 2019;**278**:424-434

[52] Gao S, Song W, Guo M. The integral role of bioproducts in the growing bioeconomy. Industrial Biotechnology. 2020;**16**(1):13-25

[53] Gupta A et al. Harnessing genetic engineering to drive economic bioproduct production in algae. Frontiers in Bioengineering and Biotechnology. 2024;**12**

[54] Sproles AE et al. Recent advancements in the genetic engineering of microalgae. Algal Research. 2021;**53**:102158

[55] Varshney P et al. Extremophilic micro-algae and their potential contribution in biotechnology. Bioresource Technology. 2015;**184**:363-372

[56] Macário IPE et al. The "bright side" of cyanobacteria: Revising the nuisance potential and prospecting innovative biotechnology-based solutions to integrate water management programs. ACS Sustainable Chemistry & Engineering. 2021;**9**(21):7182-7197

[57] Wijffels RH, Kruse O, Hellingwerf KJ. Potential of industrial biotechnology with cyanobacteria and eukaryotic microalgae. Current Opinion in Biotechnology. 2013;**24**(3):405-413

[58] Joshi S, Mishra S. Recent advances in biofuel production through metabolic engineering. Bioresource Technology. 2022;**352**:127037

[59] Ahmad A et al. Algae biotechnology for industrial wastewater treatment, bioenergy production, and high-value bioproducts. Science of the Total Environment. 2022;**806**:150585

[60] Yarkent Ç, Gürlek C, Oncel SS. Potential of microalgal compounds in trending natural cosmetics: A review. Sustainable Chemistry and Pharmacy. 2020;**17**:100304

[61] Koyande AK et al. Microalgae: A potential alternative to health supplementation for humans. Food Science and Human Wellness. 2019;**8**(1):16-24

[62] Wang X et al. A review on optimistic biorefinery products: Biofuel and bioproducts from algae biomass. Fuel. 2023;**338**:127378

[63] He Z et al. Bacterial valorization of lignin for the sustainable production of value-added bioproducts. International Journal of Biological Macromolecules. 2024;**279**:135171

[64] Jiang L, Li Y, Pei H. Algal–bacterial consortia for bioproduct generation and wastewater treatment. Renewable and Sustainable Energy Reviews. 2021;**149**:111395

[65] ÉdE S et al. Obtaining bioproducts from the studies of signals and interactions between microalgae and bacteria. Microorganisms. 2022;**10**(10):2029

[66] Zhu S et al. Microalgal consortia for waste treatment and valuable bioproducts. Energies. 2023;**16**(2):884

[67] Chio C et al. Optimization and purification of bioproducts from *Bacillus*

velezensis PhCL fermentation and their potential on industrial application and bioremediation. Science of the Total Environment. 2023;**903**:166428

[68] Peña-Montenegro TD et al. Metatranscriptomic response of deep ocean microbial populations to infusions of oil and/or synthetic chemical dispersant. Applied and Environmental Microbiology. 2024;**90**(8):e01083-e01024

[69] Passi A et al. Genome-scale metabolic modeling enables in-depth understanding of big data. Metabolites. 2022;**12**(1):14

[70] Kumar M et al. Mixotrophic growth of a ubiquitous marine diatom. Science Advances. 2024;**10**(29):eado2623

[71] Thiele I, Palsson BØ. A protocol for generating a high-quality genome-scale metabolic reconstruction. Nature Protocols. 2010;**5**(1):93-121

[72] Orth JD, Thiele I, Palsson BØ. What is flux balance analysis? Nature Biotechnology. 2010;**28**(3):245-248

[73] Canto-Encalada G et al. Ten Quick Tips for Accurate Reconstruction of Prokaryotic and Eukaryotic Genome-Scale Metabolic Models, in Preprints. 2023, Preprints

[74] Levering J et al. Genome-scale model reveals metabolic basis of biomass partitioning in a model diatom. PLoS One. 2016;**11**(5):e0155038

[75] Chang RL et al. Metabolic network reconstruction of *Chlamydomonas* offers insight into light-driven algal metabolism. Molecular Systems Biology. 2011;**7**(1):518

[76] Ye C et al. Reconstruction and analysis of the genome-scale metabolic model of *Schizochytrium limacinum* SR21 for docosahexaenoic acid production. BMC Genomics. 2015;**16**(1):799

[77] van Oossanen S, Ledo NF, dos Santos VM, Barbosa MJ, Diez MS, D'Adamo S. iSO1949_N. oceanica, Genome-scale metabolic model of Nannochloropsis oceanica. 2024

[78] Ray A, Kundu P, Ghosh A. Reconstruction of a genome-scale metabolic model of *Scenedesmus obliquus* and its application for lipid production under three trophic modes. ACS Synthetic Biology. 2023;**12**(11):3463-3481

[79] Loira N et al. Reconstruction of the microalga *Nannochloropsis salina* genome-scale metabolic model with applications to lipid production. BMC Systems Biology. 2017;**11**(1):66

[80] van Tol HM, Armbrust EV. Genome-scale metabolic model of the diatom *Thalassiosira pseudonana* highlights the importance of nitrogen and sulfur metabolism in redox balance. PLoS One. 2021;**16**(3):e0241960

[81] Shah AR et al. Reconstruction and analysis of a genome-scale metabolic model of *Nannochloropsis gaditana*. Algal Research. 2017;**26**:354-364

[82] Zuñiga C et al. Genome-scale metabolic model for the green alga *Chlorella vulgaris* UTEX 395 accurately predicts phenotypes under autotrophic, heterotrophic, and mixotrophic growth conditions. Plant Physiology. 2016;**172**(1):589-602

[83] Mekanik M et al. Improvement of lutein production in *Auxenochlorella protothecoides* using its genome-scale metabolic model and a system-oriented approach. Applied Biochemistry and Biotechnology. 2023;**195**(2):889-904

[84] Yang C, Hua Q, Shimizu K. Energetics and carbon metabolism

during growth of microalgal cells under photoautotrophic, mixotrophic and cyclic light-autotrophic/dark-heterotrophic conditions. Biochemical Engineering Journal. 2000;**6**(2):87-102

[85] León-Saiki GM et al. Metabolic modelling and energy parameter estimation of *Tetradesmus obliquus*. Algal Research. 2018;**35**:378-387

[86] Nogales J et al. Detailing the optimality of photosynthesis in cyanobacteria through systems biology analysis. Proceedings of the National Academy of Sciences. 2012;**109**(7):2678-2683

[87] Broddrick JT et al. Predicting the metabolic capabilities of *Synechococcus elongatus* PCC 7942 adapted to different light regimes. Metabolic Engineering. 2019;**52**:42-56

[88] Santos-Merino M et al. Highlighting the potential of *Synechococcus elongatus* PCC 7942 as platform to produce α-linolenic acid through an updated genome-scale metabolic modeling. Frontiers in Microbiology. 2023;**14**:1126030

[89] Klanchui A et al. An improved genome-scale metabolic model of Arthrospira platensis C1 (iAK888) and its application in glycogen overproduction. Metabolites. 2018;**8**(4):84

[90] Yoshikawa K et al. Construction of a genome-scale metabolic model of *Arthrospira platensis* NIES-39 and metabolic design for cyanobacterial bioproduction. PLoS One. 2015;**10**(12):e0144430

[91] Mueller TJ et al. Rapid construction of metabolic models for a family of cyanobacteria using a multiple source annotation workflow. BMC Systems Biology. 2013;**7**(1):142

[92] Guzmán-Palomino A et al. Dynamical simulation, sensitivity, and productivity analysis of a light-photoacclimation model for microalgae-based carbohydrate production in continuous photobioreactors. Processes. 2023;**11**(7):1866

[93] Machado D et al. Exploring the gap between dynamic and constraint-based models of metabolism. Metabolic Engineering. 2012;**14**(2):112-119

[94] Suthers PF et al. Recent advances in constraint and machine learning-based metabolic modeling by leveraging stoichiometric balances, thermodynamic feasibility and kinetic law formalisms. Metabolic Engineering. 2021;**63**:13-33

[95] Resat H, Petzold L, Pettigrew MF. Kinetic modeling of biological systems. Computational Systems Biology. 2009:311-335

[96] Lee E, Jalalizadeh M, Zhang Q. Growth kinetic models for microalgae cultivation: A review. Algal Research. 2015;**12**:497-512

[97] Aiba S. Growth kinetics of photosynthetic microorganisms. In: Microbial Reactions. 2005. pp. 85-156

[98] Han BP, Virtanen M, Koponen J, Straškraba M. Effect of photoinhibition on algal photosynthesis: A dynamic model. Journal of Plankton Research. 2000;**22**(5):865-885

[99] Papacek S, Jablonsky J, Petera K. Advanced integration of fluid dynamics and photosynthetic reaction kinetics for microalgae culture systems. BMC Systems Biology. 2018;**12**:1-2

[100] Solimeno A, Samsó R, Uggetti E, Sialve B, Steyer JP, Gabarró A, et al. New mechanistic model to simulate

microalgae growth. Algal Research. 2015;**12**:350-358

[101] Wang D, Lai YC, Karam AL, de los Reyes III FL, Ducoste JJ. Dynamic modeling of microalgae growth and lipid production under transient light and nitrogen conditions. Environmental Science & Technology. 2019;**53**(19):11560-11568

[102] Cornet JF, Dussap CG, Gros JB, Binois C, Lasseur C. A simplified monodimensional approach for modeling coupling between radiant light transfer and growth kinetics in photobioreactors. Chemical Engineering Science. 1995;**50**(9):1489-1500

[103] Krichen E, Rapaport A, Le Floc'h E, Fouilland E. A new kinetics model to predict the growth of micro-algae subjected to fluctuating availability of light. Algal Research. 2021;**58**:102362

[104] Pruvost J, Legrand J, Legentilhomme P, Muller-Feuga A. Simulation of microalgae growth in limiting light conditions: Flow effect. AIChE Journal. 2002;**48**(5):1109-1120

[105] Huang T et al. Biogeochemical Cycles. Amsterdam: Elsevier; 2023

[106] Gilbert JA et al. Meeting report: The terabase metagenomics workshop and the vision of an Earth microbiome project. Standards in Genomic Sciences. 2010;**3**:243-248

[107] Shaffer JP et al. Standardized multi-omics of Earth's microbiomes reveals microbial and metabolite diversity. Nature Microbiology. 2022;**7**(12):2128-2150

[108] Eloe-Fadrosh EA et al. The National Microbiome Data Collaborative Data Portal: An integrated multi-omics microbiome data resource. Nucleic Acids Research. 2021;**50**(D1):D828-D836

[109] Pushpakumara BLDU et al. Unravelling microalgal-bacterial interactions in aquatic ecosystems through 16S rRNA gene-based co-occurrence networks. Scientific Reports. 2023;**13**(1):2743

[110] McDonald D et al. An improved Greengenes taxonomy with explicit ranks for ecological and evolutionary analyses of bacteria and archaea. The ISME Journal. 2011;**6**(3):610-618

[111] Tibocha-Bonilla JD et al. Advances in metabolic modeling of oleaginous microalgae. Biotechnology for Biofuels. 2018;**11**(1):241

[112] Ge H, Walhout AJ, Vidal M. Integrating 'omic' information: A bridge between genomics and systems biology. Trends in Genetics. 2003;**19**(10):551-560

[113] Christopher JA et al. Subcellular proteomics. Nature Reviews Methods Primers. 2021;**1**(1):32

[114] Tardif M et al. PredAlgo: A new subcellular localization prediction tool dedicated to green algae. Molecular Biology and Evolution. 2012;**29**(12):3625-3639

[115] Li-Beisson Y et al. The lipid biochemistry of eukaryotic algae. Progress in Lipid Research. 2019;**74**:31-68

[116] Li X et al. A genome-wide algal mutant library and functional screen identifies genes required for eukaryotic photosynthesis. Nature Genetics. 2019;**51**(4):627-635

[117] Almagro Armenteros JJ et al. DeepLoc: Prediction of protein subcellular localization using

deep learning. Bioinformatics. 2017;**33**(21):3387-3395

[118] Ødum MT et al. DeepLoc 2.1: Multi-label membrane protein type prediction using protein language models. Nucleic Acids Research. 2024;**52**(W1):W215-W220

[119] Gordon DE et al. Comparative host-coronavirus protein interaction networks reveal pan-viral disease mechanisms. Science. 2020;**370**(6521):eabe9403

[120] Delseny M, Han B, Hsing YI. High throughput DNA sequencing: The new sequencing revolution. Plant Science. 2010;**179**(5):407-422

[121] Briones-Baez MF et al. Multi-objective optimization of microalgae metabolism: An evolutive algorithm based on FBA. Metabolites. 2022;**12**(7):603

[122] Flynn KJ et al. Plankton digital twins—A new research tool. Journal of Plankton Research. 2022;**44**(6):805-805

[123] Teng SY et al. Microalgae with artificial intelligence: A digitalized perspective on genetics, systems and products. Biotechnology Advances. 2020;**44**:107631

[124] Long B, Fischer B, Zeng Y, Amerigian Z, Li Q, Bryant H, et al. Machine learning-informed and synthetic biology-enabled semi-continuous algal cultivation to unleash renewable fuel productivity. Nature Communications. 2022;**13**(1):541

[125] Oruganti RK et al. Artificial intelligence and machine learning tools for high-performance microalgal wastewater treatment and algal biorefinery: A critical review. Science of the Total Environment. 2023;**876**:162797

[126] Ansari FA, Nasr M, Rawat I, Bux F. Artificial neural network and techno-economic estimation with algae-based tertiary wastewater treatment. Journal of Water Process Engineering. 2021;**40**:101761

Chapter 8

Biocompounds of Commercial Interest from Freshwater and Marine Phytoplankton

Irene Gallego

Abstract

Microalgae are photosynthetic microorganisms that produce a wide range of biocompounds, such as proteins, omega-3 fatty acids or pigments like astaxanthin; with various applications in the pharmaceutical, cosmetic, bioenergy and food sectors. This chapter provides an overview of the compounds and molecules synthesized by microalgae, ranging from polysaccharides to vitamins, minerals and other secondary metabolites. Additionally, the chapter reviews the key biological aspects that influence the production of such biocompounds, including strain selection, strain improvement and cultivation conditions, as well as the biotechnological factors necessary to optimize the production and processing of these compounds, such as cultivation system, extraction and purification. Lastly, the chapter presents the main applications of commercially relevant microalgae-derived compounds, emphasizing the most notable microalgae-based products currently being developed in the global market.

Keywords: microalgae, bioactive compounds, nutraceuticals, biofuels, animal feed, food supplements, cosmetics, biotechnology

1. Introduction

Biocompounds —chemical compounds derived from biological sources— are gaining attention in the global market because of their sustainable production and higher acceptance compared to synthetic compounds. In 2010, one third of the total β-carotene production had a 'natural' origin [1]. Today, the high demand for omega-3 fatty acids is searching for alternative, non-fish sources to reach the expected market size of 3.8 billion USD (US dollars) by 2026 [2].

Microalgae, a diverse group of photosynthetic microorganisms responsible for capturing 50% of the global atmospheric carbon [3], can also synthetize dozens of biocompounds with various commercial applications, including the pharmaceutical sector, food and feed industries, and the bioenergy sector, among others [4]. These cell microfactories use sunlight and carbon dioxide (CO_2) to sustainably produce high-value biocompounds, such as proteins, lipids, polysaccharides, pigments, vitamins, minerals and other secondary metabolites [5]. The commercial exploitation of some microalgae-derived biocompounds, such as β-carotene, astaxanthin, omega-3 fatty acids, pigments and extracts, is well stablished [1], with a global market size valued in 3.4 billion USD in 2020 [6]. Some microalgal biocompounds, with relatively low production costs, have been commercialized for decades. This is the case of β-carotene from *Dunaliella salina*, with a market price that may reach 300–1500 USD/kg, commercialized since the 1980s [1]. Other biocompounds, however, are still far from being economically viable and require further biotechnological advancements.

Microalgae are predominantly single-celled organisms widely distributed in aquatic ecosystems, —including freshwater, marine or brackish environments— as well as in terrestrial habitats, such as soils and sand [4]. Due to their fast life cycles and short generation times (around 24 h), microalgae can generate relatively high biomass without the need of fertile soils, as opposed to terrestrial crops [7].

According to Whittaker's classification system [8], microalgae can be divided into eukaryotic microalgae and prokaryotic cyanobacteria. While organisms belonging to the former group contain membrane-bound organelles, such as the nucleus or the chloroplast, prokaryotes lack these structures. Within the eukaryotic microalgae group, the most abundant groups belong to phylum Chlorophyta (green algae), phylum Ochrophyta (class Bacillariophyceae includes diatoms and class Chrysophyceae includes golden algae) and phylum Miozoa (class Dinophyceae includes dinoflagellates) [9]. Prokaryotic cyanobacteria (phylum Cyanobacteria) are commonly referred to as blue-green algae. Interestingly, taxonomists use the term 'microalgae' to refer exclusively to the eukaryotic microalgae. However, the physiology and biotechnological applications of both groups of microorganisms are similar, and cyanobacteria are colloquially considered as microalgae [7]. In this chapter, for the sake of simplicity, I will use the term 'microalgae' to indistinctively refer to cyanobacteria and/or eukaryotic microalgae.

Microalgae only require a few essential resources to grow and reproduce, namely light and nutrients. In most cases, a liquid culture medium with nitrogen, phosphorus and silica, in the case of diatoms will suffice to allow microalgal growth. However, to achieve competitive productivities and an economically viable production of microalgal biomass —and derived biocompounds—, some factors must be considered, such as the strain and optimal growth parameters, the type of cultivation system and the bioprocessing techniques to obtain the final product. Before targeting the production of one (or more) biocompounds of interest, it is recommended to follow a roadmap to clarify some questions, such as: (1) which strain(s) produce(s) larger amounts of biocompound(s); (2) how to optimize the strain cultivation conditions (nutrients, environmental variables, genetic techniques); (3) what is the preferred cultivation system (open systems *vs*. closed systems); and (4) what are the most common downstream processing techniques (harvesting and processing). In this chapter, I will focus on the most relevant biocompounds and metabolites that are produced by microalgae, how to increase their production yields, and what are the applications of these valuable compounds that are currently under development, or available in the market.

2. Compounds of interest

2.1 Main biocompounds of interest produced by microalgae

Microalgae are an excellent source of primary metabolites that are necessary for their own survival, e.g., proteins, essential amino acids, lipids, fatty acids, etc.; and secondary metabolites, which are not essential for microalgae survival, and include hydrocarbons, pigments and vitamins, among others (see **Figure 1**) [10].

2.1.1 Proteins and derivatives

The two most common microalgae cultivated nowadays, *Chlorella* spp. (Chlorophyta) and spirulina—commercial name for *Arthrospira*/*Limnospira* spp. (phylum Cyanobacteria) [11]—have been historically exploited as an alternative source of proteins, as their average composition may contain up to 60–70% of proteins (dry weight, DW) [12]. The first documented records of the consumption of microalgae go back to the sixteenth century, when the Spanish chroniclers described that the Aztecs harvested spirulina from Mexican lagoons [13]. *Chlorella* spp. also has a relatively long tradition of culturing and consumption in Asia due to its high protein content, and commercial cultivation at large scale started to develop in Japan in the 1960s [13]. *Chlorella* spp. and spirulina are an excellent source of proteins in plant-based diets, generally with higher concentrations of essential amino acids (not synthetized) than other plant or animal protein sources [14]. Additionally, microalgae-derived peptides have demonstrated antioxidant, immune-protector, anticancer, hepatoprotective and anticoagulant properties, but the technology readiness level to produce microalgal peptides for biomedical applications is still in its early stage [15].

Microalgae also produce enzymes that can be directly used in the industry, and enzymes that are involved in the biosynthesis of microalgal biocompounds (carotenoids, peptides, etc.). For instance, L-asparaginase is an enzyme widely used in the food processing industry and is mainly produced not only by bacteria, but also by cyanobacteria, such as *Limnospira maxima* [16].

Mycosporine-like amino acids (MAAs) are a group of ~40 secondary metabolites that have recently gained commercial interest because of their photoprotective function against ultraviolet (UV) radiation [17, 18], being commonly referred to as

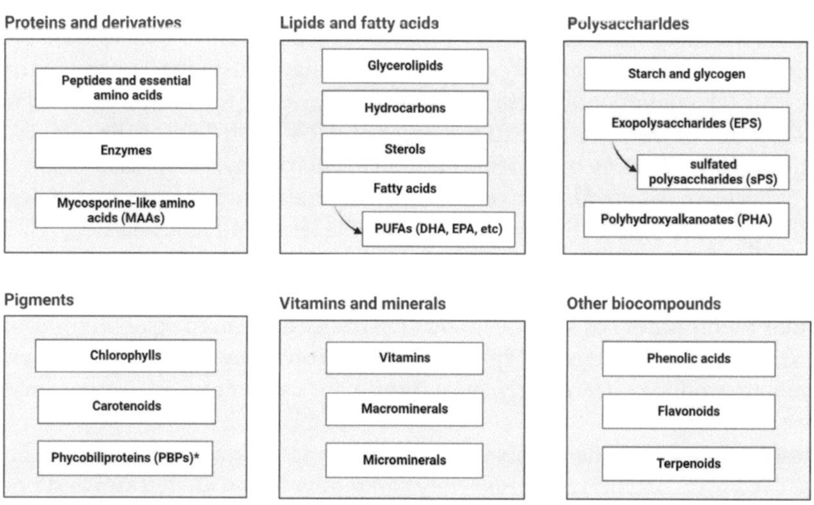

Figure 1.
*Compounds of commercial interest produced by microalgae, classified into six main groups. Primary metabolites are given in white boxes and secondary metabolites are given in green boxes. Arrows point to different subtypes of compounds. *Phycobiliproteins (PBPs), composed by proteins and phycobilins, are shown as pigments.*

'microbial sunscreens'. More than 150 species of marine microalgae are known to produce MAAs, including cyanobacteria, green microalgae, red microalgae, diatoms, cryptophyceans and dinophyceans [19].

Phycobiliproteins (PBPs) are proteins bound to microalgal photosynthetic pigments (phycobilins) that capture and transfer light inside the cell. Examples of PBPs are further detailed in the specific section on pigments (Section 2.1.4).

2.1.2 Lipids and fatty acids

Microalgae are renowned for their high ability to accumulate lipids, in comparison with plant oil crops [20]. The yield of microalgal oils is strain dependent, and several species have received high attention for commercial exploitation as potential biofuels and nutrition supplements. For example, the unicellular marine thraustochytrid *Aurantiochytrium* (formerly *Schizochytrium,* phylum Bigyra) may reach up to 77% of lipid content inside the cell, and the green microalgae *Botryococcus braunii* (Chlorophyta) can store up to 75% of lipids. Lipid accumulation is also dependent on the cultivation conditions, and nitrogen limitation has shown to be very effective to increase the lipid content [21].

Microalgae produce a large range of lipid-like compounds, such as glycerolipids, sterols and hydrocarbons (see also **Figure 1**) [22]. Fatty acids (FAs), the building blocks of lipids, can be categorized into two main groups: neutral, e.g., triacylglycerols (TAGs); or polar, with a more complex structure, e.g., long-chain polyunsaturated fatty acids (LC-PUFAs) [23].

Microalgal PUFAs have attracted commercial interest because of their multiple health benefits (antioxidant, anti-inflammatory, antibacterial activities and a protective effect against cardiovascular problems) [24]. The most extensively studied PUFAs are omega-3 docosahexaenoic acid (DHA), eicosapentaenoic acid (EPA) and arachidonic acid (ARA). Because mammals are unable to synthesize these molecules, they must be acquired through the diet. Traditionally, the primary source for PUFAs is fish oil, but its quality and availability are unfortunately declining due to pollution and unsustainable management practices. Microalgae are seen as a promising alternative and sustainable source of PUFAs, as they produce at least 70 different PUFAs [25]. *Aurantiochytrium* sp. and *Ulkenia* sp. (Bigyra) have the highest productivity of DHA [26]. The red microalga *Porphyridium purpureum* (formerly *Porphyridium cruentum*) is one of the richest natural sources of ARA, with 36% of the total FA composition [14]. *Phaeodactylum tricornutum* (Bacillariophyceae), *Pavlova* spp. (Pavlovophyceae), *Nannochloropsis oculata* (Eustigmatophyceae) and *Isochrysis galbana* (Coccolithophyceae) have also shown considerable levels of DHA and EPA [27, 28].

Glycerolipids (or glycolipids) are lipids with a glycerol backbone and FAs and mono- or oligosaccharide attached, and they can have either a structural function (in cellular membranes) or storage lipids (TAGs). Green microalgae (*Tetraselmis* sp., *Scenedesmus* sp., *Tetradesmus* spp.) and the diatom *Phaeodactylum tricornutum* are known to produce different glycolipids with anticancer and anti-inflammatory activities [29].

Phytosterols are steroids widely used as additives in many food products. They have received great attention because they can reduce blood cholesterol and prevent cardiovascular disorders [29]. *Diacronema lutheri* (formerly *Pavlova lutheri*) (Pavlovophyceae) is the most promising strain for microalgal phytosterols, with a content over 5% DW [30].

Hydrocarbons derived from fatty acids, i.e., alkanes and alkenes, are ubiquitous not only in plants, in insects' external tissues, but also in cyanobacteria. Alka(e)nes of different chain lengths are important targets for biotechnology because they are major components of gasoline, jet fuels and diesel fuels. For instance, *Chlamydomonas reinhardtii, Chlorella variabilis* (Chlorophyta) and *Microchloropsis gaditana* (formerly *Nannochloropsis gaditana*) (Eustigmatophyceae) produce relatively large amounts of heptadecene [31]. Other relevant hydrocarbons for commercial use include carotenes, a subtype of carotenoid that will be described in Section 2.1.4.

2.1.3 Polysaccharides

Microalgae polysaccharides (and oligosaccharides) are considered as byproducts of pigments and/or lipids production [32], but there are some polysaccharides with industrial uses, e.g., moisturizing and aggregating agents in cosmetics, or substrates for bioethanol production [33, 34].

The most relevant polysaccharides for industrial exploitation include exopolysaccharides (EPS), i.e., saccharides excreted outside the cells as mucilage or in the culture media. The advantage of EPS is that the downstream processing (extraction and purification steps) can be simpler than for intracellular compounds [32]. Other polysaccharides of commercial interest are starch and glycogen, which can be used as feedstock and chemical products (**Figure 1**).

The composition of EPS may include sugars (glucose, galactose, fucose, xylose, arabinose, rhamnose, mannose, fructose, etc.) and other non-sugar substituents, such as proteins, or sulfated groups. Both freshwater and marine microalgae (mainly green microalgae, diatoms) segregate EPS with a content of protein and sulfate that can reach up to 26% of the total EPS composition (up 20% sulfate groups and 9% proteins). For more details, see [35, 36] and references therein.

Sulfated polysaccharides (sPS) have drawn attention because of their pharmaceutical and biomedical application; with antiviral, anti-inflammatory, antioxidant, hypoglycemic and anticoagulant properties [35]. For instance, the segregated sPS from *Porphyridium purpureum* (red microalga) show a strong inhibitory effect on *Herpes simplex* virus [37]. Spirulan, another sPS extracted from spirulina, presents antibiotic bioactivity [38]. Additionally, sPS can be used for other industrial applications, as drag reducers and biolubricants (see review in Ref. [33]).

Eukaryotic microalgae and cyanobacteria use the polysaccharides such as starch and glycogen, respectively, as energy storage. Microalgal starch is a potential candidate to replace synthetic polymers, particularly those used in packaging. Some examples of microalgae that produce a high content of intracellular starch are *Chlorella vulgaris*, *Tetraselmis subcordiformis*, *Chromochloris zofingiensis* and *Parachlorella kessleri* (Chlorophyta) [39]. Moreover, glycogen derived from cyanobacteria has demonstrated potential as an alternative fermentation feedstock to produce liquid fuels and chemicals. Heterotrophic organisms, e.g., yeasts, can use glycogen from cyanobacteria as substrate to produce biofuel [40]. The glycogen content in cyanobacteria is dependent on the strain and the growth conditions. For instance, while *Synechococcus* sp. PCC 6803 has shown a content of glycogen equivalent to 12% (DW) [41], spirulina may contain 13.7–63% of glycogen (DW), depending on the cultivation conditions [42].

Cyanobacteria also have the ability to autotrophically produce polyhydroxyalkanoates (PHAs) as storage compounds. PHAs are biopolymers (polyesters) produced by bacteria and cyanobacteria, and serve as a promising alternative to petroleum-based plastics. A recent review on cyanobacterial PHA production identified only six strains producing five types of PHAs relevant for the chemical industry, concluding that PHA productivity is still far from being economically viable and that research should expand towards more screening studies and genetic-engineered strains [43].

2.1.4 Pigments

Microalgae may contain three different pigments: chlorophylls, carotenoids and phycobiliproteins (**Figure 1**).

Chlorophylls are green pigments present in photosynthetic organisms (from microalgae and cyanobacteria to higher plants), involved in light energy absorption during photosynthesis, and the synthesis of other pigments [44]. In microalgae, the estimated amount of chlorophyll is around 1% DW [45]. Chlorophylls are commercially exploited as food colorants, cosmetic ingredients, and in biopharma due to their coloring properties, stimulating effects and antioxidant action [44]. There are five major groups of chlorophylls based on their colors and light absorption characteristics. The most abundant types of chlorophyll are chlorophyll *a* (dark green), particularly abundant in cyanobacteria and red microalgae —but present in all microalgae— and chlorophyll *b* (brilliant green), present in green microalgae. Spirulina is a good source of chlorophyll *a*, while *Chlorella* spp. contains high concentrations of both chlorophylls *a* and *b* [46]. Chlorophylls *c*, *d* and *f* are less abundant and are only present in certain microalgal groups [47]. Extraction of chlorophylls can be challenging, since the pigments can become unstable under oxygen, light, temperature and pH variations [45, 48].

Carotenoids are colored, liposoluble pigments that are responsible for the yellow, orange, red or purple color of some microalgae, as well as plants (fruits, vegetables and flowers), to protect them against photodamage. At least 1204 natural carotenoids have been described from 722 source organisms, 297 of which are algae (macro- and microalgae) [49]. Carotenoids can be classified into two groups: Carotenes, which are hydrocarbons, and xanthophylls, which are oxygenated derivatives of carotenes. Generally, the most commonly exploited carotene worldwide is β-carotene, while the most commonly used xanthophylls are astaxanthin, lutein, fucoxanthin and zeaxanthin. Carotenoids account for an average of 0.1% of the DW of algae, but some microalgae may reach 14% under certain growth conditions, such as the halophytic *Dunaliella salina* (Chlorophyta) [50]. Due to their photoprotective properties, carotenoids are widely used as antioxidants for human health and well-being.

Astaxanthin is the most stable carotenoid and has the most potent antioxidant action [51]. *Haematococcus lacustris* (formerly *Haematococcus pluvialis*) (Chlorophyta) can naturally accumulate up to 5% astaxanthin (DW), but achieving economically viable yields of astaxanthin is challenging and requires careful consideration of numerous factors [52]. *Chromochloris zofingiensis* can achieve productivity of astaxanthin close to 2.0 mg/L/d [53].

β-Carotene is a red-orange carotenoid, precursor of vitamin A (retinol), and together with lutein, it is one of the most pigments with the highest market value. The economic demand for β-carotene is expected to increase 3.8% yearly (from 2018 to 2026), with an expected global market size of 620 million USD [54]. *Dunaliella salina* (*Dunaliella bardawil*) are the most important species that naturally accumulate β-carotene (14%, DW) [55], with a usual range of β-carotene concentrations between 0.1 and 1 mg/L, in large production systems [56].

Lutein is also a potent antioxidant and has drawn much interest in its health-promoting functions. Dosages of 6 mg day^{-1} have been proven to be beneficial for human health, including eye health [57]. The lutein market is also important and may reach 357.7 million USD by the end of 2024 [58]. Several microalgae, including species of the genera *Chlorella*, *Chlamydomonas*, *Desmodesmus*, *Dunaliella* and *Scenedesmus*, naturally accumulate lutein in a range of 4.5–7.05 mg/g (DW) [59, 60], with dozens of reported cultivation strategies (even under different cultivation modes) to increase the content of lutein, particularly for *Chlorella sorokiniana* and *Auxenochlorella protothecoides* (formerly *Chlorella protothecoides*) (Chlorophyta) (see review in [61]).

Fucoxanthin, a brown pigment mostly present in marine microalgae, accounts for more than 10% of the total natural carotenoids [46]. Although the main natural sources of fucoxanthin are brown macroalgae, many microalgae groups (e.g., diatoms, haptophytes, chrysophytes, etc.) are potential candidates for fucoxanthin production. In addition to the antioxidant effect, fucoxanthin has also reported anticancer properties (antiproliferation of cancer cells and cytotoxicity) [62]. *Isochrysis galbana*, *Isochrysis zhanjiangensis* and *Tisochrysis lutea* (Coccolithophyceae) may contain up to 23 mg/g fucoxanthin (DW) after optimization [57]. The diatoms *Phaeodactylum tricornutum*, *Cylindrotheca* sp., *Odontella aurita*, *Chaetoceros muelleri*, *Amphora* sp. and *Navicula* sp. also show considerable fucoxanthin production capacity (see review in Ref. [63]).

Zeaxanthin and canthaxanthin are the orange-color xanthophylls, both precursors of astaxanthin. Both pigments have antioxidant properties and are used as food colorants [64]. Red microalgae (*P. purpureum* and *Rhodosorus* sp.), green microalgae *Chloroidium ellipsoideum* (formerly *Chlorella ellipsoidea*) and cyanobacteria (*Arthrospira* spp., *Synechococcus* sp.) are the main sources of microalgal zeaxanthin [57, 65, 66], while canthaxanthin is naturally occurring in *Chlorella vulgaris* and *Coelastrella striolata* (Chlorophyta) [65]. The production of these pigments requires complex downstream processing (extraction and separation), which increases costs [57].

Phycobiliproteins are water-soluble fluorescent pigments, commonly present in cyanobacteria, red microalgae and some cryptophytes. Attending the pigment colors and light absorption characteristics, there are four groups of PBPs, namely, phycocyanin (PC), allophycocyanin (APC), phycoerythrin (PE) and phycoerythrocyanin (PEC). PC is a blue pigment abundant in spirulina, *Aphanizomenon* and other cyanobacteria; APC is also a blue pigment present in extremophilic red microalga *Galdieria sulphuraria* (Rhodophyta, Cyanidiophytina); PE is a red pigment from *Porphyridium* spp. and *Rhodomonas salina* (Cryptophyceae), and PEC is a magenta pigment found only in some cyanobacteria [67]. PBPs are currently used as natural colorants in cosmetics and the food industry. Moreover, they show antioxidant and anti-inflammatory properties and can be used as fluorescent markers [68–70].

2.1.5 Vitamins and minerals

Microalgae constitute an alternative source of vitamins and minerals that can fulfill human and animal nutritional requirements. Consumption of microalgae-derived vitamins may offer some advantages over other sources: renewable source, low-carbon footprint production, good absorption, vegan origin and no toxic byproducts, and higher vitamin content than some terrestrial plants [71, 72]. Besides the high concentration of β-carotene (precursor of vitamin A), microalgae also contain vitamins C, D, E and K, and the entire B group. In their seminal study, Fabrega and Herrero [73] demonstrated that four marine microalgae (*Tetraselmis suecica*, *Isochrysis galbana*, *Dunaliella tertiolecta*, *Chlorella stigmatophora*) contain higher concentrations of vitamins A, E, B_1 and B_9 compared to conventional food sources typically recognized for their vitamin content. Another study showed that *Diacronema lutheri* (formerly *Pavlova lutheri*; Pavlovophyceae), *Tetraselmis suecica* (Chlorophyta) and *Skeletonema costatum* (Mediophyceae) contain ~20-fold more vitamin D than cod liver oil [74]. Vitamin K_1, essential for prevention of chronic diseases, was found in the highest concentrations in *Anabaena cylindrica* (Cyanobacteria), with levels of 200 µg/g (DW) [75].

Minerals represent the inorganic portion of human and animal diets, and can be classified as macronutrients (e.g., calcium, phosphorus, potassium, sodium, magnesium, sulfur) or micronutrients (e.g., zinc, iron, copper, manganese, cobalt). The mineral composition of microalgae is not only heterogeneous and varies across species, but also indicates variation between marine and freshwater microalgae [76]. A study that evaluated the mineral content of 11 microalgal strains revealed that marine *Tetraselmis chui* (Chlorophyta) contained significantly higher levels of calcium, copper and zinc than other strains; *Phaeodactylum tricornutum* and *Porphyridium aerugineum* (also marine species) were richest in magnesium and iron, respectively; and freshwater *Botryococcus braunii* contained the highest levels of phosphorus and manganese [77].

2.1.6 Other secondary metabolites

Microalgae produce other valuable compounds to different industries and human health, such as phenolic acids and flavonoids. These compounds with antioxidant properties have gained attention recently, but their bioavailability and bioefficacy are limited [78]. Cyanobacteria (*Anabaena* spp., *Phormidium* sp., *Nostoc* sp.) and Chlorophyta (*Scenedesmus* spp., *Chlorella* sp., *Haematococcus* sp.) may produce up to 20 different phenolic acids and flavonoids [79].

Terpenes (squalene, pinene, limonene, bisabolene) are isoprenoids with aromatic properties, responsible for scents. Due to the low production, metabolic engineering is currently applied to increase terpenoids' yields [80].

2.2 Factors influencing the productivity of compounds of interest

There are four main factors that must be carefully considered to enhance microalgal biomass production or microalgal biocompounds of interest: the type of strain, cultivation parameters, cultivation mode and cultivation system.

2.2.1 Strain selection

The selection of the strain will mainly depend on the product of interest for commercial use. Strains can be purchased from microalgae culture collections, donated by laboratories or directly isolated from the natural environment. The advantages of microalgae culture collections (living libraries of biological resources) are various: the strains have been previously isolated and are ready to use, additional information is provided (genetics, morphological and physiological traits, culture conditions), and usually meet regulatory standards. Strain isolation is a time-consuming process that involves different techniques, including traditional techniques such as serial dilution and/or micro-pipetting, and/or automated techniques, such as flow cytometry with cell sorting [81].

Additionally, purification methods must be applied if the ultimate goal is to obtain axenic strains. In this case, techniques such as sonication, agar plating and the use of antibiotics, might be necessary to decontaminate the microalgal culture [14, 81].

Strain growth rate is in some way interconnected to cell/colony size [82]. Strains of larger sizes are expected to show lower growth rates than smaller size strains, but a likely trade-off might arise, i.e., bigger strains might contain higher concentrations of biocompounds, in comparison with smaller strains. Therefore, using only strain size as a parameter for strain selection would not be recommended, as it is not informative enough.

Strain selection can be challenging due to the unknown productivity of the target compound(s), or because our target strain might be (physiologically) similar to other strains. In this case, strain screening is essential prior to strain selection. Since microalgae screening studies can include from dozens to hundreds of strains, e.g., [83, 84], the use of bioinformatics and AI tools can facilitate the microalgal analysis by providing more accurate identification and quantification of the strains [85].

2.2.2 Strain genomic improvement

To increase production yields, microalgae can be genetically improved by selective breeding, either with random (UV-induced) or with targeted (gene editing) mutagenesis. Microalgal production systems can also be optimized using genetic and metabolic engineering, by adding (or removing) specific genes in the metabolic pathways involved in the production and accumulation of biocompounds [86].

2.2.3 Growth parameters

Choosing the most appropriate growth culture medium to ensure long-term cultivation is critical to avoid microalgae stress and to achieve a competitive productivity, at a later stage [81]. Furthermore, resource levels (nitrogen, phosphorus and light are essential non-substitutable resources for microalgae) and environmental conditions (pH, temperature, aeration, CO_2 levels) are highly connected to microalgal productivities and biomass [14, 87]. Optimization of growth conditions for each strain and product of interest remains crucial for enhancing productivity. A simple search of the terms 'microalga*' AND (('growth condition*' OR 'growth parameter*' OR 'optim*')) in the bibliographic database Web of Science shows more than 10,700 peer-reviewed scientific publications [88], which gives an estimation of the relevance of these factors to maximize the microalgae production. Since analyzing the output of Web of Science is beyond the scope of this chapter, I will only describe briefly the importance of light, temperature, pH, nutrient starvation, aeration and CO_2 levels.

It is well established that light intensity and the duration of the light:dark cycle (photoperiod) have a strong influence on microalgal growth, and that high light intensities, above the maximum threshold, will cause photoinhibition [14, 87]. Also, different light wavelengths can induce the production of certain compounds. For instance, it is known that microalgal lipid production is enhanced under relatively high light intensities (but the amount of PUFAs might decrease), and that the production of certain carotenoids is stimulated by blue light wavelengths [89, 90].

Temperature also plays a crucial role in cell growth and metabolism, and each strain is likely to have a different (negatively skewed hump-shaped) thermal performance curve, with optimal temperature levels relatively close to the threshold values, as it occurs in natural populations of phytoplankton [91]. The amplitude of the thermal gradient is also dependent on the strain, which is usually wider (and reach higher optimal temperatures) for cyanobacteria, and narrower (with lower optimal temperatures) for diatoms [92].

Nutrient starvation is recognized as one of the most common strategies to increase the production of storage compounds [14, 81]. In general, nitrogen limitation enhances the intracellular concentration of lipids or saccharides. A recent review on microalgal lipid production showed that 117 out of 189 studies induced nitrogen starvation [93]. Also, it is well known that the synthesis and accumulation of carotenoids, such as β-carotene and astaxanthin, can be stimulated under low nitrogen conditions [14]. In the case of diatoms (with silicon as a non-substitutable resource), low levels of silicon rather than low nitrogen, enhance lipid accumulation [14]. Increasing the salinity concentration in the medium is another strategy to induce the intracellular accumulation of valuable compounds. High salinity induces the accumulation of flavonoids, probably due to their antioxidant activity [79], and has also proven to be a highly effective method for enhancing lipid content [94].

Maintaining an equilibrium between air and dissolved CO_2 in the culture is crucial for the optimal strain growth. CO_2 removal rates increase with increasing biomass densities, and consequently, pH and dissolved CO_2 will fluctuate. Fluctuations in pH and dissolved CO_2 might change the availability of metals and minerals such as iron and calcium, and essential nutrients such as phosphorus [95]. These fluctuations can (and must) be minimized with CO_2 injections and/or continuous aeration. Also, pH is strain dependent, and strains from the same genus can show totally different values. For instance, *Chlorella vulgaris* optimal growth occurs at pH values between 7.5 and 8.0 [96], but *Chlorella sorokiniana* grows better at pH 6 [97].

The key variable(s) for enhancing the production of high-value biocompounds will depend on both the strain and type of biocompound. A recent review (>200 studies) on growth variables from 95 marine and freshwater phytoplankton species revealed that temperature, light-dark cycle and irradiance levels were the most frequently manipulated parameters to enhance the production of lipids, concluding that understudied factors, such as pH, phosphorus limitation or metals, might lead to higher lipid yields [93].

2.2.4 Cultivation mode

Three cultivation modes can be used for large-scale microalgae production: photoautotrophic (using CO_2 and light to generate biomass), mixotrophic (using an organic carbon source in the presence of light) and heterotrophic (using an organic carbon source in the absence of light) conditions. Mixotrophic cultivation allows shifting from photoautotrophy (in the presence of light and CO_2) to heterotrophy under dark conditions, while heterotrophic cultivation cannot use CO_2 as a carbon source, since the accumulation of the gas will decrease the pH of the medium. Mixotrophic cultures have demonstrated higher biomass productivity than photoautotrophic cultures, and more cost-efficient system than heterotrophic cultivation modes. The use of byproducts as organic substrates (e.g., acetate) enhances biomass productivity and reduces cost and environmental impacts [98].

2.2.5 Cultivation system

Microalgae can be cultivated in open or closed systems. Open ponds (OPs) are shallow ponds or tanks that usually allow mixing the culture, have low energy requirements and construction costs, and are easy to scale up. Photobioreactors (PBRs) are closed systems (transparent culture vessels) designed to enhance light penetration and photosynthesis. Fermenters are also closed systems used to cultivate mixotrophic and heterotrophic microalgae. The advantage of closed systems is that growth parameters can be easily controlled and monitored, there is low risk of contamination and the scale-up process is accelerated. PBRs are the predominant cultivation systems within the European Union (EU) [99]. However, the global use of fermenters has increased over the last decade, particularly for food production [100].

A recent meta-analysis on cultivation modes (OP *vs* PBRs) elucidated that the environmental performance of microalgae cultivation not only depends on the cultivation system, but also on the location and the species considered, concluding that no cultivation system is favorable in terms of productivity [101]. OP systems require the least amount of land because these systems are usually placed in locations with high temperatures and irradiances, which increase the productivity [101].

2.2.6 Downstream processing

Extraction and purification of microalgal compounds is challenging because conventional techniques are highly energy-dependent. Common harvesting techniques include mechanical pressing, milling and solvent extraction, which are time- and energy-costly [102]. Indeed, ionic liquids may be toxic and pose environmental risks if they are not treated properly before discharge [103]. The use of genetic techniques to produce cell-wall deficient microalgal strains can be advantageous to optimize the production system, but in general, these techniques have a focus on improving a microalgal trait with a clear commercial application rather than improving the production system [104]. Sometimes, strains that were improved with random-mutagenesis techniques are used instead. This is the case of the model organism *Chlamydomonas reinhardtii* and its recombinant cell-wall deficient UVM4 strain.

2.3 Commercial applications of microalgae

2.3.1 Dietary supplements and functional foods

The global market of dietary supplements and functional foods has exponentially increased in the last years, both in terms of sales and variety of products available [105]. Microalgae play an important role in both sectors due to the variety of valuable compounds including pigments with antioxidant properties, PUFAs and vitamins, and the high protein content and balanced amino acid profiles, with numerous health benefits [15, 106]. A bioinformatics-based review on microalgae genomes and metagenomes shows that microalgae contain compounds with antiviral, antibacterial, anti-inflammatory, anticancerogenic, antioxidant, immune-protective and prebiotic activity [107]. Microalgal extracts rich in astaxanthin, β-carotene, lutein or chlorophylls, soft gels containing microalgal oils rich in omega-3 PUFAs, lyophilized capsules containing microalgal biomass rich in vitamins and proteins, are some examples of what is currently available in the market [108].

Also, feed enriched with microalgal biomass improves animal immune response, disease resistance and gut function of the animal. Camacho et al. reviewed the health benefits of microalgae-derived biocompounds on animals, and they proposed potential industrial applications of microalgae to increase the quality of cattle, poultry, piglets, lamb, fish and crustaceans' meat, as well as the production (and quality) of eggs [109]. For example, astaxanthin is utilized in aquaculture as feed additive, not only to enhance the color of farmed fish and shrimp, but also to improve the quality of seafood for human consumption [110].

Incorporating microalgae into foods can also lead to potential benefits for human and animal health due to the presence of several biocompounds. Nowadays, it is not uncommon to find in our supermarkets and stores a variety of products with microalgal extracts as ingredients: biscuits, breads, snacks, oils, drinks, pasta, condiments, emulsions and dairy products [15, 106, 111]. A higher consumer demand of non-animal protein sources and more sustainable production might explain this trend.

Additionally, most microalgal pigments (chlorophylls, β-carotene, astaxanthin, lutein, phycocyanin, phycoerythrin) are used in the food and beverage industry as natural food/beverage colorants, to replace synthetic colorants and ensure food safety. For example, phycocyanin from spirulina is a blue natural colorant commonly used in beverages and some foods.

Spirulina and chlorella, colloquially referred to as 'superfoods', dominate the microalgal market due to their nutrient-rich profile: high protein content, PUFAs, pigments, vitamins (including the B group) and minerals [12].

2.3.2 Fertilizers and biostimulants

The new paradigm of circular bioeconomy, with valorization of the microalgal residual biomass after extraction of high-value compounds, has potential applications in agriculture. Crop production and quality can be improved with the use of microalgae-derived biostimulants rich in minerals and micronutrients. Some benefits of using microalgae include the mineralization with simpler molecules for direct uptake by plants, plant protection against pathogens, pH buffering, higher resilience against stressors (droughts or salt stress) and stimulation of plant growth [112]. Osorio-Reyes et al. compiled some recent applications of foliar and soil application of microalgae and their effects on plant crops, and spirulina and/or *Chlorella* spp. were the most common biostimulants [113].

2.3.3 Cosmetics and personal care products

Different biocompounds are used in the cosmetics industry. Astaxanthin and lutein are usually included in cosmetics formulations due to their powerful antioxidants and antiaging activity [52]. MAAs are well-stablished photoprotective agents, used in sunscreens, anti-photoaging agents and wound-healing agents [18]. EPS is endowed with moisturizing and hydrating properties [32].

2.3.4 Biofuels

Despite the considerable amount of research on biofuels over the past decades, large-scale production of third- and fourth-generation biofuels (with microalgae as feedstock) remains challenging, but the microalgal biofuel industry (particularly in China) is expected to experience rapid development [114]. Alternative bioenergy sources include the production of biohydrogen, bioethanol or biobutanol, among others [115]. Genetically improved cyanobacteria can produce polysaccharides that will be used as a substrate for the production of bioethanol (by anaerobic fermentation) or produce bioethanol directly [86].

2.3.5 Pharmaceuticals and biomedicine

Microalgae can also be genetically transformed to generate cell factories with applications in biomedicine, such as the production of recombinant proteins. A recent review on microalgal recombinant vaccines has compiled 18 vaccines against various animal and human infectious diseases, with *Chlamydomonas reinhardtii* as the preferred cell factory [116].

2.3.6 Biopolymers

The production of microalgal biopolymers, such as PHAs, starch and cellulose, may contribute to a more sustainable economy and reduce the production of non-biodegradable plastics. Nonetheless, the industrial production of microalgal polymers is still in its infancy, but the number of scientific studies on microalgae-derived plastics is growing [117].

2.3.7 Bioremediation and wastewater treatment

Microalgae can remove nutrients, metals and organic contaminants, and have been proven to decontaminate industrial, urban and agricultural effluents. Wastewater treatment plants use microalgae in their tertiary treatment to remove pollutants and excess macronutrients -nitrogen (N), phosphorus (P)- from effluents. Green microalgae (*Tetradesmus obliquus, Chlorella* spp., *Chlamydomonas* spp., filamentous *Spirogyra* spp.) and cyanobacteria (*Limnospira maxima*) have a high nutrient and/or metal removal rate (see review in Ref. [118]). Microalgae can also improve air quality by removing toxic gases from the atmosphere and increasing oxygen concentrations, as shown in Ref. [119].

2.3.8 Carbon sequestration

Microalgae are known as the most efficient biological sequesters of CO_2, which are used to produce compounds of interest. The voluntary carbon markets are including microalgae production as a nature-based solution to reduce the emission of greenhouse gases, and the number of microalgal biotechnologies providing this service has recently multiplied [108].

2.4 Examples of microalgae-based products currently commercialized

Microalgae are commercially exploited worldwide. Here, I selected some examples of microalgal producers across the globe (**Table 1**). In the Pacific region, the Hawaiian biotech Cyanotech Corporation is one of the pioneers in microalgae nutritional supplements, and a world leader in astaxanthin production from *Haematococcus lacustris* [120], followed by the Swedish AstaReal. In Japan, with a long tradition of incorporating microalgae as food ingredients, the biotech corporation Euglena Co. has developed a full line of products from *Euglena gracilis* (Euglenophyta) as functional food and nutritional supplement. The omega-3 fatty acid DHA is produced at industrial scale in at least 20 countries, including France [121]. DHA from *Schizochytrium* sp. is generally recognized as safe (GRAS) by the US Food and Drug Administration (FDA) and commonly added to infant formula products. In Inner Mongolia (China), the largest agglomeration of alkaline lakes allows the cultivation of spirulina under optimal growth conditions, with an annual production estimated in >3000 t [122].

Location	Main compound	Strain	Cultivation system	Product/service	Notes
Hawaii (USA)	Astaxanthin Spirulina biomass	*H. lacustris*	OP, PBR	BioAstin® nutritional supplement	Net sales (2024) 15.1 M USD
Nacka (Sweden)	Astaxanthin	*H. lacustris*	Indoor bioreactor	AstaReal®	Estimated revenue (2024) 16.9 M USD
Tokyo (Japan)	Biomass rich in amino acids and glutamic acid	*Euglena gracilis*		"Euglena for the Body" nutritional supplement and health foods	Net sales (2024) 23.649 M JPY (2024)
Libourne (France)	LC-PUFAs (DHA)	*Schizochytrium* sp.	Fermenter	DHA ORIGINS©	—
Inner Mongolia (China)	Spirulina biomass Phycocyanin	Spirulina	OP (enclosed)	Food supplements Food colorants (blue)	87 Ha cultivation area Annual production >3000 Ton
Malaysia (HQ in Japan)	Biomass	—	Flat-panel PBR	Carbon capture	Largest carbon capture farm (5 Ha, 100 Ha in 3 years)
Lyon (France)	Recombinant proteins	Genetically improved strains	PBR, fermenter	NINKARAK® and ALGAVAX® platforms	—

Abbreviations: HQ = headquarters, OP = open pond, PBR = photobioreactor, H. lacustris = Haematococcus lacustris.

Table 1.
List of seven examples of relevant microalgae-derived products currently commercialized or under development, at a global scale.

Innovative services include carbon sequestration using microalgae to boost the corporate climate finance (voluntary carbon markets). The largest microalga biomass production farm for carbon capture is located in Malaysia, with an estimated total surface of 100 Ha in 2027. In France, microalgal cell chassis are genetically improved to produce therapeutic recombinant proteins at large scale: Immunotoxins for immunotherapy, thermostable vaccines, etc.

3. Conclusions

Microalgae production is increasing worldwide due to the growing interest in biocompounds. These microorganisms are a potent source of proteins, lipids, pigments, vitamins and other valuable compounds with industrially relevant applications. Different economic sectors, including the food and feed industries, agriculture, cosmetics, pharmaceutical and biomedicine, bioenergy sector, or carbon capture markets, are incorporating microalgal compounds to fulfill the society demands and to achieve a more sustainable economy. Selection of the adequate strain and optimization of the growth conditions (e.g., light intensity and photoperiod, temperature, nutrient optimal and limiting levels, salinity, pH) are crucial steps to maximize the production of microalgae valuable compounds. Yet, productivities of some microalgal-derived compounds are not fully optimized, but new screening techniques, innovative cultivation and bioprocessing technologies, as well as strain improvement with genetic and metabolic engineering, may contribute to enhance the production of microalgal biocompound(s) and to move towards a circular bioeconomy.

Conflict of interest

The author declares no conflict of interest.

Author details

Irene Gallego
Research Division Agroecology and Environment, Agroscope, Zurich, Switzerland

*Address all correspondence to: irene.gallego@agroscope.admin.ch

References

[1] Borowitzka MA. High-value products from microalgae—Their development and commercialisation. Journal of Applied Phycology. 2013;**25**(3):743-756

[2] Magoni C, Bertacchi S, Giustra CM, Guzzetti L, Cozza R, Ferrari M, et al. Could microalgae be a strategic choice for responding to the demand for omega-3 fatty acids? A European perspective. Trends in Food Science & Technology. 2022;**121**:142-155

[3] Field CB, Behrenfeld MJ, Randerson JT, Falkowski P. Primary production of the biosphere: Integrating terrestrial and oceanic components. Science. 1998;**281**:237-241

[4] Metting FB Jr. Biodiversity and application of microalgae. Journal of Industrial Microbiology. 1996;**17**:477-489

[5] Levasseur W, Perre P, Pozzobon V. A review of high value-added molecules production by microalgae in light of the classification. Biotechnology Advances. 2020;**41**:107545

[6] Loke SP. Global market and economic analysis of microalgae technology: Status and perspectives. Bioresource Technology. 2022;**357**:127329

[7] Thore ESJ, Muylaert K, Bertram MG, Brodin T. Microalgae. Current Biology. 2023;**33**(3):R91-RR5

[8] Whittaker RH. New concepts of kingdoms or organisms. Evolutionary relations are better represented by new classifications than by the traditional two kingdoms. Science. 1969;**163**(3863):150-160

[9] Ruggiero MA, Gordon DP, Orrell TM, Bailly N, Bourgoin T, Brusca RC, et al. A higher level classification of all living organisms. PLoS One. 2015;**10**(4):e0119248

[10] Maplestone RA, Stone MJ, Williams DH. The evolutionary role of secondary metabolites — A review. Gene. 1992;**115**:151-157. Available from: Maplestone_1992_Gene_REV_secondary_metabolites.pdf

[11] Nowicka-Krawczyk P, Muhlsteinova R, Hauer T. Detailed characterization of the Arthrospira type species separating commercially grown taxa into the new genus Limnospira (cyanobacteria). Scientific Reports. 2019;**9**(1):694

[12] Abreu AP, Martins R, Nunes J. Emerging applications of Chlorella sp. and Spirulina (Arthrospira) sp. Bioengineering. 2023;**10**(8):955-989

[13] FAO Fisheries and Aquaculture Circular No. 1034. Rome: Food and Agriculture Organization of the United Nations; 2008. 41 p

[14] Richmond A. Handbook of Microalgal Culture: Biotechnology and Applied Phycology. Iowa: Blackwell Publishing Company; 2004. 566 p. DOI: 10.1002/9781118567166

[15] Caporgno MP, Mathys A. Trends in microalgae incorporation into innovative food products with potential health benefits. Frontiers in Nutrition. 2018;**5**:58

[16] El Baky HHA, El Baroty GS, Mostafa EM. Optimization growth of Spirulina (Arthrospira) Platensis in Photobioreactor under varied nitrogen concentration for maximized biomass, carotenoids and lipid contents. Recent Patents on Food, Nutrition & Agriculture. 2020;**11**(1):40-48

[17] Chrapusta E, Kaminski A, Duchnik K, Bober B, Adamski M, Bialczyk J. Mycosporine-like amino acids: Potential health and beauty ingredients. Marine Drugs. 2017;**15**(10):326-355

[18] Görünmek M, Ballık B, Çakmak ZE, Çakmak T. Mycosporine-like amino acids in microalgae and cyanobacteria: Biosynthesis, diversity, and applications in biotechnology. Algal Research. 2024;**80**:103507

[19] Llewellyn CA, Airs RL. Distribution and abundance of MAAs in 33 species of microalgae across 13 classes. Marine Drugs. 2010;**8**(4):1273-1291

[20] Udayan A, Pandey AK, Sirohi R, Sreekumar N, Sang BI, Sim SJ, et al. Production of microalgae with high lipid content and their potential as sources of nutraceuticals. Phytochemistry Reviews. 2023;**22**:833-860

[21] Griffiths MJ, van Hille RP, Harrison STL. Lipid productivity, settling potential and fatty acid profile of 11 microalgal species grown under nitrogen replete and limited conditions. Journal of Applied Phycology. 2011;**24**(5):989-1001

[22] Fernandez-Valenzuela S, Chávez-Ruvalcaba F, Beltran-Rocha JC, San Claudio PM, Reyna-Martínez R. Isolation and culturing axenic microalgae: Mini-review. The Open Microbiology Journal. 2021;**15**(1):111-119

[23] Guiheneuf F, Stengel DB. LC-PUFA-enriched oil production by microalgae: Accumulation of lipid and triacylglycerols containing n-3 LC-PUFA is triggered by nitrogen limitation and inorganic carbon availability in the marine haptophyte Pavlova lutheri. Marine Drugs. 2013;**11**(11):4246-4266

[24] Archer L, Mc Gee D, Paskuliakova A, McCoy GR, Smyth T, Gillespie E, et al. Fatty acid profiling of new Irish microalgal isolates producing the high-value metabolites EPA and DHA. Algal Research. 2019;**44**:101671

[25] Maltsev Y, Maltseva K. Fatty acids of microalgae: Diversity and applications. Reviews in Environmental Science and Bio/Technology. 2021;**20**(2):515-547

[26] Aasen IM, Ertesvag H, Heggeset TM, Liu B, Brautaset T, Vadstein O, et al. Thraustochytrids as production organisms for docosahexaenoic acid (DHA), squalene, and carotenoids. Applied Microbiology and Biotechnology. 2016;**100**(10):4309-4321

[27] Adarme-Vega TC, Lim DK, Timmins M, Vernen F, Li Y, Schenk PM. Microalgal biofactories: A promising approach towards sustainable omega-3 fatty acid production. Microbial Cell Factories. 2012;**11**:96-106. Available from: Adarme-Vega_2012_DHA-EPA.pdf

[28] Aussant J, Guiheneuf F, Stengel DB. Impact of temperature on fatty acid composition and nutritional value in eight species of microalgae. Applied Microbiology and Biotechnology. 2018;**102**(12):5279-5297

[29] Saide A, Martinez KA, Ianora A, Lauritano C. Unlocking the health potential of microalgae as sustainable sources of bioactive compounds. International Journal of Molecular Sciences. 2021;**22**:4383-4423

[30] Ahmed F, Zhou W, Schenk PM. Pavlova lutheri is a high-level producer of phytosterols. Algal Research. 2015;**10**:210-217

[31] Sorigue D, Legeret B, Cuine S, Morales P, Mirabella B, Guedeney G, et al. Microalgae synthesize hydrocarbons from long-chain fatty acids via a light-dependent pathway. Plant Physiology. 2016;**171**(4):2393-2405

[32] Pierre G, Delattre C, Dubessay P, Jubeau S, Vialleix C, Cadoret JP, et al. What is in store for EPS microalgae in the next decade? Molecules. 2019;**24**:4296-4321

[33] Xiao R, Zheng Y. Overview of microalgal extracellular polymeric substances (EPS) and their applications. Biotechnology Advances. 2016;**34**(7):1225-1244

[34] Karemore A, Sen R. Downstream processing of microalgal feedstock for lipid and carbohydrate in a biorefinery concept: A holistic approach for biofuel applications†. RSC Advances. 2016;**6**(35):29486-29496

[35] Raposo MF, de Morais RM, Bernardo de Morais AM. Bioactivity and applications of sulphated polysaccharides from marine microalgae. Marine Drugs. 2013;**11**(1):233-252

[36] Morais MG, Santos TD, Moraes L, Vaz BS, Morais EG, Costa JAV. Exopolysaccharides from microalgae: Production in a biorefinery framework and potential applications. Bioresource Technology Reports. 2022;**18**:101006

[37] Huleihel M, Ishanu V, Tal J, Arad S. Antiviral effect of red microalgal polysaccharides on herpes simplex and varicella zoster viruses. Journal of Applied Phycology. 2001;**13**(2):127-134

[38] Hayashi K, Hayashi T, Kojima I. A natural Sulfated polysaccharide, calcium Spirulan, isolated from Spirulina platensis: In vitro and ex vivo evaluation of anti-herpes simplex virus and anti-human immunodeficiency virus activities. AIDS Research and Human Retroviruses. 1996;**12**(15):1463-1471

[39] Noguchi M, Aizawa R, Nakazawa D, Hakumura Y, Furuhashi Y, Yang S, et al. Application of real treated wastewater to starch production by microalgae: Potential effect of nutrients and microbial contamination. Biochemical Engineering Journal. 2021;**169**:107973

[40] Möllers KB, Cannella D, Jørgensen H, Frigaard N-U. Cyanobacterial biomass as carbohydrate and nutrient feedstock for bioethanol production by yeast fermentation. Biotechnology for Biofuels. 2014;**7**(1):64

[41] De Porcellinis A, Frigaard NU, Sakuragi Y. Determination of the glycogen content in cyanobacteria. Journal of Visualized Experiments. 2017;**125**:e56068

[42] Hasunuma T, Kikuyama F, Matsuda M, Aikawa S, Izumi Y, Kondo A. Dynamic metabolic profiling of cyanobacterial glycogen biosynthesis under conditions of nitrate depletion. Journal of Experimental Botany. 2013;**64**(10):2943-2954

[43] Rueda E, Gonzalez-Flo E, Mondal S, Forchhammer K, Arias DM, Ludwig K, et al. Challenges, progress, and future perspectives for cyanobacterial polyhydroxyalkanoate production. Reviews in Environmental Science and Bio/Technology. 2024;**23**(2):321-350

[44] Sun D, Wu S, Li X, Ge B, Zhou C, Yan X, et al. The structure, functions and potential medicinal effects of chlorophylls derived from microalgae. Marine Drugs. 2024;**22**:65-83

[45] Hosikian A, Lim S, Halim R, Danquah MK. Chlorophyll extraction from microalgae: A review on the process engineering aspects. International Journal of Chemical Engineering. 2010;**2010**:1-11

[46] Sun H, Wang Y, He Y, Liu B, Mou H, Chen F, et al. Microalgae-derived pigments for the food industry. Marine Drugs. 2023;**21**:82-109

[47] Queiroz Zepka L, Jacob-Lopes E, Roca M. Catabolism and bioactive properties of chlorophylls. Current Opinion in Food Science. 2019;**26**:94-100

[48] Ezhumalai G, Arun M, Manavalan A, Rajkumar R, Heese K. A holistic approach to circular bioeconomy through the sustainable utilization of microalgal biomass for biofuel and other value-added products. Microbial Ecology. 2024;**87**(1):61

[49] Carotenoids Database. 2020. Available from: http://carotenoiddb.jp

[50] Chen Z, Wu W, Wen Y, Zhang L, Wu Y, Farid MS, et al. Recent advances of natural pigments from algae. Food Production, Processing and Nutrition. 2023;**5**:39-53

[51] Dufossé L, Galaup P, Yaron A, Arad SM, Blanc P, Chidambara Murthy KN, et al. Microorganisms and microalgae as sources of pigments for food use: A scientific oddity or an industrial reality? Trends in Food Science & Technology. 2005;**16**(9):389-406

[52] Patel AK, Tambat VS, Chen CW, Chauhan AS, Kumar P, Vadrale AP, et al. Recent advancements in astaxanthin production from microalgae: A review. Bioresource Technology. 2022;**364**:128030

[53] Sun H, Ren Y, Lao Y, Li X, Chen F. A novel fed-batch strategy enhances lipid and astaxanthin productivity without compromising biomass of Chromochloris zofingiensis. Bioresource Technology. 2020;**308**:123306

[54] Silva SC, Ferreira ICFR, Dias MM, Barreiro MF. Microalgae-derived pigments: A 10-year Bibliometric review and industry and market trend analysis. Molecules. 2020;**25**(15):3406

[55] Aasen AJ, Eimhjellen KE, Liaaen-Jensen S. An extreme source of β-carotene. Acta Chemica Scandinavica. 1969;**23**:2544-2545. Available from: Aasen_1969_ActaChemScand_beta-carotene.pdf

[56] Pourkarimi S, Hallajisani A, Alizadehdakhel A, Nouralishahi A, Golzary A. Factors affecting production of beta-carotene from Dunaliella salina microalgae. Biocatalysis and Agricultural Biotechnology. 2020;**29**:101771

[57] Ren Y, Sun H, Deng J, Huang J, Chen F. Carotenoid production from microalgae: Biosynthesis, salinity responses and novel biotechnologies. Marine Drugs. 2021;**19**:713-734

[58] Liu C, Hu B, Cheng Y, Guo Y, Yao W, Qian H. Carotenoids from fungi and microalgae: A review on their recent production, extraction, and developments. Bioresource Technology. 2021;**337**:125398

[59] Razzak SA. Comprehensive overview of microalgae-derived carotenoids and their applications in diverse industries. Algal Research. 2024;**78**:103422

[60] Christaki E, Bonos E, Giannenas I, Florou-Paneri P. Functional properties of carotenoids originating from algae. Journal of the Science of Food and Agriculture. 2013;**93**(1):5-11

[61] Coleman B, Vereecke E, Van Laere K, Novoveska L, Robbens J. Genetic engineering and innovative cultivation strategies for enhancing the lutein production in microalgae. Marine Drugs. 2024;**22**:329-353

[62] Kumar SR, Hosokawa M, Miyashita K. Fucoxanthin: A marine carotenoid exerting anti-cancer effects by affecting multiple mechanisms. Marine Drugs. 2013;**11**(12):5130-5147

[63] Khaw YS, Yusoff FM, Tan HT, Noor Mazli NAI, Nazarudin MF, Shaharuddin NA, et al. Fucoxanthin

production of microalgae under different culture factors: A systematic review. Marine Drugs. 2022;**20**:592-617

[64] Rebelo BA, Farrona S, Ventura MR, Abranches R. Canthaxanthin, a red-hot carotenoid: Applications, synthesis, and biosynthetic evolution. Plants (Basel). 2020;**9**:1039-1057

[65] Raposo MF, de Morais AM, de Morais RM. Carotenoids from marine microalgae: A valuable natural source for the prevention of chronic diseases. Marine Drugs. 2015;**13**(8):5128-5155

[66] Bourdon L, Jensen AA, Kavanagh JM, McClure DD. Microalgal production of zeaxanthin. Algal Research. 2021;**55**:102266

[67] Bryant DA. Phycoerythrocyanin and phycoerythrin: Properties and occurrence in Cyanobacteria. Journal of General Microbiology. 1982;**128**:835-844. Available from: Bryant_1982_JGenMic_phycoerythrocyanin.pdf

[68] Hamed I. The evolution and versatility of microalgal biotechnology: A review. Comprehensive Reviews in Food Science and Food Safety. 2016;**15**(6):1104-1123

[69] Garcia JL, de Vicente M, Galan B. Microalgae, old sustainable food and fashion nutraceuticals. Microbial Biotechnology. 2017;**10**(5):1017-1024

[70] Wang J, Qin S, Lin J, Wang Q, Li W, Gao Y. Phycobiliproteins from microalgae: Research progress in sustainable production and extraction processes. Biotechnology for Biofuels and Bioproducts. 2023;**16**(1):170

[71] Del Mondo A, Smerilli A, Sane E, Sansone C, Brunet C. Challenging microalgal vitamins for human health. Microbial Cell Factories. 2020;**19**(1):201

[72] Arora Y, Sharma S, Sharma V. Microalgae in bioplastic production: A comprehensive review. Arabian Journal for Science and Engineering. 2023;**48**(6):7225-7241

[73] Fabregas J, Herrero C. Vitamin content of four marine microalgae. Potential use as source of vitamins in nutrition. Journal of Industrial Microbiology. 1990;5:259-264. Available from: Fabregas_Herrero_JIndMic_1990_Vitamins.pdf

[74] De Roeck-Holtzhauer Y, Quere I, Claire C. Vitamin analysis of five planktonic microalgae and one macroalga. Journal of Applied Phycology. 1991;**3**:259-264. Available from: De Roeck-Holtzhauer_1991_vitamins.pdf

[75] Tarento TDC, McClure DD, Vasiljevski E, Schindeler A, Dehghani F, Kavanagh JM. Microalgae as a source of vitamin K1. Algal Research. 2018;**36**:77-87

[76] Fox JM, Zimba PV. Minerals and trace elements in microalgae. In: Microalgae in Health and Disease Prevention. London: Elsevier; 2018. pp. 177-193. DOI: 10.1016/B978-0-12-811405-6.00008-6

[77] Tibbetts SM, Milley JE, Lall SP. Chemical composition and nutritional properties of freshwater and marine microalgal biomass cultured in photobioreactors. Journal of Applied Phycology. 2014;**27**(3):1109-1119

[78] Bouayed J, Hoffmann L, Bohn T. Total phenolics, flavonoids, anthocyanins and antioxidant activity following simulated gastro-intestinal digestion and dialysis of apple varieties: Bioaccessibility and potential uptake. Food Chemistry. 2011;**128**(1):14-21

[79] Del Mondo A, Smerilli A, Ambrosino L, Albini A, Noonan DM, Sansone C, et al. Insights into phenolic compounds from microalgae: Structural

variety and complex beneficial activities from health to nutraceutics. Critical Reviews in Biotechnology. 2021;**41**(2):155-171

[80] Huang PW, Wang LR, Geng SS, Ye C, Sun XM, Huang H. Strategies for enhancing terpenoids accumulation in microalgae. Applied Microbiology and Biotechnology. 2021;**105**(12):4919-4930

[81] Andersen RA. Algal Culturing Techniques. New York: Elsevier Academic Press; 2005. 578 p

[82] Kremer CT, Thomas MK, Litchman E. Temperature- and size-scaling of phytoplankton population growth rates: Reconciling the Eppley curve and the metabolic theory of ecology. Limnology and Oceanography. 2017;**62**(4):1658-1670

[83] Rodolfi L, Chini Zittelli G, Bassi N, Padovani G, Biondi N, Bonini G, et al. Microalgae for oil: Strain selection, induction of lipid synthesis and outdoor mass cultivation in a low-cost photobioreactor. Biotechnology and Bioengineering. 2009;**102**(1):100-112

[84] Lee K, Eisterhold ML, Rindi F, Palanisami S, Nam PK. Isolation and screening of microalgae from natural habitats in the midwestern United States of America for biomass and biodiesel sources. Journal of Natural Science, Biology and Medicine. 2014;**5**(2):333-339

[85] Alzahmi AS, Daakour S, Nelson D, Al-Khairy D, Twizere J-C, Salehi-Ashtiani K. Enhancing algal production strategies: Strain selection, AI-informed cultivation, and mutagenesis. Frontiers in Sustainable Food Systems. 2024;**8**:1331251

[86] Gupta A, Kang K, Pathania R, Saxton L, Saucedo B, Malik A, et al. Harnessing genetic engineering to drive economic bioproduct production in algae. Frontiers in Bioengineering and Biotechnology. 2024;**12**:1350722

[87] Daneshvar E, Sik Ok Y, Tavakoli S, Sarkar B, Shaheen SM, Hong H, et al. Insights into upstream processing of microalgae: A review. Bioresource Technology. 2021;**329**:124870

[88] Web of Science. 29 October 2024. Available from: www.webofscience.com

[89] Maltsev Y, Maltseva K, Kulikovskiy M, Maltseva S. Influence of light conditions on microalgae growth and content of lipids, carotenoids, and fatty acid composition. Biology (Basel). 2021;**10**:1060-1084

[90] Zhang Z, Han T, Sui J, Wang H. Cryptochrome-mediated blue-light signal contributes to carotenoids biosynthesis in microalgae. Frontiers in Microbiology. 2022;**13**:1083387

[91] Litchman E, Klausmeier CA. Trait-based community ecology of phytoplankton. Annual Review of Ecology, Evolution, and Systematics. 2008;**39**(1):615-639

[92] Rossi S, Carecci D, Ficara E. Thermal response analysis and compilation of cardinal temperatures for 424 strains of microalgae, cyanobacteria, diatoms and other species. Science of the Total Environment. 2023;**873**:162275

[93] Morales M, Aflalo C, Bernard O. Microalgal lipids: A review of lipids potential and quantification for 95 phytoplankton species. Biomass and Bioenergy. 2021;**150**:106108

[94] Suparmaniam U, Lam MK, Lim JW, Yusup S, Tan IS, Lau SY, et al. Influence of environmental stress on microalgae growth and lipid profile: A systematic review. Phytochemistry Reviews. 2023;**22**(4):879-901

[95] Gerardi. Algae, alkalinity, and pH. In: The Biology and Troubleshooting of Facultative Lagoons. New Jersey: Wiley; 2015. pp. 105-109. DOI: 10.1002/9781118981771

[96] Rachlin JW, Grosso A. The effects of pH on the growth of Chlorella vulgaris and its interactions with cadmium toxicity. Archives of Environmental Contamination and Toxicology. 1991;**20**:505-508. Available from: Rachlin_Grosso_1991_pH_Cvulgaris.pdf

[97] Qiu R, Gao S, Lopez PA, Ogden KL. Effects of pH on cell growth, lipid production and CO2 addition of microalgae Chlorella sorokiniana. Algal Research. 2017;**28**:192-199

[98] Proietti Tocca G, Agostino V, Menin B, Tommasi T, Fino D, Di Caprio F. Mixotrophic and heterotrophic growth of microalgae using acetate from different production processes. Reviews in Environmental Science and Bio/Technology. 2024;**23**(1):93-132

[99] Araújo R, Vázquez Calderón F, Sánchez López J, Azevedo IC, Bruhn A, Fluch S, et al. Current status of the algae production industry in Europe: An emerging sector of the blue bioeconomy. Frontiers in Marine Science. 2021;**7**:626389

[100] Garofalo C, Norici A, Mollo L, Osimani A, Aquilanti L. Fermentation of microalgal biomass for innovative food production. Microorganisms. 2022;**10**:2069-2089

[101] Schade S, Meier T. A comparative analysis of the environmental impacts of cultivating microalgae in different production systems and climatic zones: A systematic review and meta-analysis. Algal Research. 2019;**40**:101485

[102] Balasubramaniam V, Gunasegavan RD, Mustar S, Lee JC, Mohd Noh MF. Isolation of industrial important bioactive compounds from microalgae. Molecules. 2021;**26**:943-988

[103] Tan JS, Lee SY, Chew KW, Lam MK, Lim JW, Ho SH, et al. A review on microalgae cultivation and harvesting, and their biomass extraction processing using ionic liquids. Bioengineered. 2020;**11**(1):116-129

[104] Gallego IR. Assessing the environmental impact of gene-edited microalgae

[105] Lam M, Khoshkhat P, Chamani M, Shahsavari S, Dorkoosh FA, Rajabi A, et al. In-depth multidisciplinary review of the usage, manufacturing, regulations & market of dietary supplements. Journal of Drug Delivery Science and Technology. 2022;**67**:102985

[106] Koyande AK, Chew KW, Rambabu K, Tao Y, Chu D-T, Show P-L. Microalgae: A potential alternative to health supplementation for humans. Food Science and Human Wellness. 2019;**8**(1):16-24

[107] Krohn I, Menanteau-Ledouble S, Hageskal G, Astafyeva Y, Jouannais P, Nielsen JL, et al. Health benefits of microalgae and their microbiomes. Microbial Biotechnology. 2022;**15**(7):1966-1983

[108] Gallego I, Medic N, Pedersen JS, Ramasamy PK, Robbens J, Vereecke E, Romeis J. The microalgal sector in Europe: Towards a sustainable bioeconomy. New Biotechnology. 2024. (in press)

[109] Camacho F, Macedo A, Malcata F. Potential industrial applications and commercialization of microalgae in the functional food and feed industries: A short review. Marine Drugs. 2019;**17**:312-337

[110] Lim KC, Yusoff FM, Shariff M, Kamarudin MS. Astaxanthin as feed supplement in aquatic animals. Reviews in Aquaculture. 2018;**10**(3):738-773

[111] Ampofo J, Abbey L. Microalgae: Bioactive composition, health benefits, safety and prospects as potential high-value ingredients for the functional food industry. Food. 2022;**11**:1744-1764

[112] Nichols K. Microalgae as a Beneficial Soil Amendment [White Paper]. MyLand Company LLC; 2020;**1**:1:22. Available from: Nichols_2020_white_paper_soil_amendment.pdf

[113] Osorio-Reyes JG, Valenzuela-Amaro HM, Pizana-Aranda JJP, Ramirez-Gamboa D, Melendez-Sanchez ER, Lopez-Arellanes ME, et al. Microalgae-based biotechnology as alternative biofertilizers for soil enhancement and carbon footprint reduction: Advantages and implications. Marine Drugs. 2023;**21**:93-117

[114] Wang M, Ye X, Bi H, Shen Z. Microalgae biofuels: Illuminating the path to a sustainable future amidst challenges and opportunities. Biotechnology for Biofuels and Bioproducts. 2024;**17**(1):10

[115] Cavelius P, Engelhart-Straub S, Mehlmer N, Lercher J, Awad D, Bruck T. The potential of biofuels from first to fourth generation. PLoS Biology. 2023;**21**(3):e3002063

[116] Ramos-Vega A, Angulo C, Bañuelos-Hernández B, Monreal-Escalante E. Microalgae-made vaccines against infectious diseases. Algal Research. 2021;**58**:102408

[117] Madadi R, Maljaee H, Serafim LS, Ventura SPM. Microalgae as contributors to produce biopolymers. Marine Drugs. 2021;**19**:466-493

[118] Pacheco D, Rocha AC, Pereira L, Verdelhos T. Microalgae water bioremediation: Trends and hot topics. Applied Sciences. 2020;**10**:1886-1902

[119] Barati B, Fazeli Zafar F, Amani Babadi A, Hao C, Qian L, Wang S, et al. Microalgae as a natural CO2 sequester: A study on effect of tobacco smoke on two microalgae biochemical responses. Frontiers in Energy Research. 2022;**10**:881758

[120] Cyanotech Corporation 2024 Annual Report. Available from: https://www.cyanotech.com/wp-content/uploads/2024/07/2024-Cyanotech-Annual-Report.pdf

[121] Finco AMO, Mamani LDG, Carvalho JC, de Melo Pereira GV, Thomaz-Soccol V, Soccol CR. Technological trends and market perspectives for production of microbial oils rich in omega-3. Critical Reviews in Biotechnology. 2017;**37**(5):656-671

[122] Chen J, Wang Y, Benemann JR, Zhang X, Hu H, Qin S. Microalgal industry in China: Challenges and prospects. Journal of Applied Phycology. 2015;**28**(2):715-725